**Dyes and Dyeing**

WAID ACADEMY

ANSTRUTHER

SESSION 1978-1979

# FIRST PRIZE

AWARDED TO

*STELLA AUCHINLECK*

**CLASS**
*Senior*

*(Art)*

**RECTOR**

**MAX
SIMMONS**

# Dyes and Dyeing

VAN NOSTRAND REINHOLD AUSTRALIA
Melbourne
New York
London
Toronto

Published in 1978 by
Van Nostrand Reinhold Australia Pty Ltd
17 Queen Street, Mitcham, Victoria 3132
Van Nostrand Reinhold Company
135 West 50th Street, New York, N.Y. 10020
Van Nostrand Reinhold Ltd
1410 Birchmount Road, Scarborough, Ontario
Van Nostrand Reinhold Company Ltd
Molly Millar's Lane, Wokingham, Berkshire

© 1978 Van Nostrand Reinhold

Designed by Vane Lindesay
Typeset by Savage & Co. Pty Ltd
Printed in Singapore by Toppan Printing Company

National Library of Australia Card Number
and ISBN 0 442 25016 9

This book is sold subject to the condition that it
shall not, by way of trade, be lent, resold, hired out,
or otherwise disposed of without the publisher's
consent, in any form of binding or cover other than
that in which it is published.

All rights reserved. No part of this work covered by the
Copyright hereon may be reproduced or used in any form or
by any means—graphic, electronic, or mechanical,
including photocopying, recording, taping, or information
storage systems—without written permission of the
publisher.

# Acknowledgments

The author wishes to thank the Melbourne College of Textiles for assistance in preparing some of the photographs in this book, and in particular, John Noble for his valued advice.

*Photographic credits:*
Tony Boyd: pages 7, 8, 9, 16, 18, 20, 26 (figure 13), 32, 65.
L. Bennie: pages 14, 15, 26 (figure 14), 49, 51, 54, 56, 58, 67, 68, 69.

# Contents

| | | |
|---|---|---|
| | Introduction | 1 |
| 1 | Basic Concepts | 2 |
| 2 | Natural and man-made fibres | 12 |
| 3 | Scouring and Bleaching | 19 |
| 4 | Methods of application for commercial dyes | 25 |
| 5 | The mordant process for the natural dyestuffs | 41 |
| 6 | Dyeing wool and cotton with natural products | 48 |
| 7 | Special Effects | 64 |
| 8 | Tanning recipes for sheep skins | 72 |
| | Appendix: Chemical and Dyestuff Supply Sources | 78 |
| | Index: Chemicals and auxiliaries | 80 |
| | General | 83 |

# Introduction

The art of dyeing is very much like that of cooking. Both outcomes are transmitted via the senses to the brain, and the recipient of these stimuli in one case, senses the effect by the taste buds, the other through the optics of the eye.

In both instances the recipe is the instruction needed to ensure a successful reproduction of an effect at some time later. The dyer and the cook can take delight in a well presented final product, and the ensuing job satisfaction is manifest at a successful conclusion.

One does not have to be a good cook to be a successful dyer, but the strategems' used are similar. Attention to detail, a well ordered approach to the job, and having the appropriate utensils are common to the two arts.

As in cooking, there are Cordon Bleu dyers who do not slavishly follow methods or recipes of others. Knowing the basic principles and foundations of their trade, they can experiment into the unknown areas to find the ultimate masterpieces.

I will take you through those principles and foundations so that you may experiment for yourself, to gradually refine your techniques and to produce something that you will be proud to exhibit. You will then be an individual who can be counted among the Craft people.

# 1 Basic Concepts

**Dyeing Theory**

Most organic fibrous matter is capable of absorbing water. If that water contains a colouring matter either in a molecular or colloidal state of dispersion and the solution is brought in contact with say wool, cotton, nylon, terylene or orlon, preferably at an elevated temperature, it frequently happens that most of the colouring matter leaves the water solution and becomes attached to the fibrous materials. The colour intensity of the solution gradually diminishes as is to be expected, while that of the fibre gradually increases. The fibres are said to have been dyed, the process involved is termed the dyeing process, and the solution called the dye bath will have become exhausted as the transference of its dye content to the fibre takes place.

Usually only certain members of chemical classes of dyes can be satisfactorily applied from simple water solutions to a specific type of fibre and thus there will be a class of dyes that can be applied to wool, another to cotton and so on. Whereas most dye classes will stain all fibres, the dyeing operation, as distinct from the staining operation, means fibres coloured with correct fastness to light, washing etc.

As a general rule, other compounds termed assistants, are used in the dyebath in order to produce fast and evenly coloured fibres, and in some cases the proper fixation and/or development can take place only in the presence of certain metal irons derived from chemical salts. This latter process is called mordanting, and can be applied to the fibre either before dyeing, after or concurrently.

The most important aim of dyeing is the colouration of fibres in such a way that they exhibit uniform, i.e. level, effects in predetermined shades of specified degrees of fastness to various agencies.

The dyeing operation will bring you into contact with many new terms which are necessary if you are to carry out the dyeing process in a logical manner. To understand the contents of this book you do not need a doctorate in chemistry. You will be introduced to the new terms gradually to help you through the calculations.

The main areas discussed in this chapter are

Temperature
Weights and volumes
Liquor Ratio
Percentage calculations
Concentration calculations
Acidity and alkalinity
Dyeing vessel construction
Glassware
Safety with chemicals

## Temperature

All temperatures are given in degrees Celsius. (°C) Where the recipe or instructions tell you to start the dyeing cold, ordinary cold tap water can be used. Where a definite temperature is specified, do not worry about adjusting the bath too finely as temperatures that vary ±5°C will make no difference to the outcome. Remember that when the recipe calls for a boil, the dyebath should be boiling and having reached this temperature, the heat source can be adjusted to give a gentle simmer. You will be instructed at various sections to bring the dyebath to the boil over a period, say 40 minutes. Therefore you should adjust the rate of temperature increase so that the bath reaches the boil in a smooth even manner. You can purchase a laboratory thermometer graduated from 0°C to 100°C from most chemical supply houses.

## Weights and volumes

All weights will be shown in the metric system, grams (g) or kilograms (kg). Most recipes in the following pages are based on a material weight of 50 g and in some cases 5 g, so that if you are dyeing more or less than this amount you should scale up or down accordingly.

All volumes are shown in millilitres (ml) or in litres.

*Conversions*
1000 grams = 1 kilogram
1000 ml = 1 litre

## Liquor ratio

This figure is quoted to indicate how much water is needed to prepare the dyebath. It is the ratio between the weight of water needed, to the weight of material to be dyed. A liquor ratio of 30:1 would indicate that 30 times more water than material is needed to prepare the bath. As most recipes call for a 30 : 1 L.R. and we are dyeing, say 50 g of textile material, the volume of water needed would be calculated as

$$30 \times 50 = 1500 \text{ ml or } 1 \cdot 5 \text{ litres}$$

For a 5 g sample at a L.R. of 30:1

$$30 \times 5 = 150 \text{ ml}$$

## Percentage calculations

All dyes and most chemicals needed in the dyeing operation are calculated on the weight of material to be dyed. To prepare a recipe we add x% of red dye, y% of yellow dye and maybe z% of a blue dye. Throughout the early stages of the book, all calculations will be done for you and providing you are dyeing the same weight of material as shown in the examples, you will not have to calculate for yourself. However, as the formula is so simple and relies only on basic arithmetic, you should practice the two forms of the Standard Formula. The first form is where we are calculating amounts of *dry powder* whether it be dye or chemicals, and where we are using chemicals that have not been broken down in strength or diluted. The second form can be used where we have diluted the dyes or chemicals.

*Standard Formula* (for undiluted dyes and chemicals)

$$\frac{\text{Percentage required}}{100} \times \frac{\text{Weight of material}}{1} = \text{Weight of chemical needed for dyeing.}$$

Assume the recipe calls for 3% Ammonia to be used on 50 g of wool.

Weight of Ammonia needed $= \frac{3}{100} \times \frac{50}{1} = 1\cdot5$ ml.

*Standard Formula* (for diluted dyes and chemicals)

For the home dyer it is much easier to prepare the dye powders and chemicals in solution form with hot water and allocate a known volume of this prepared solution to the dyebath. In this way most dyes and chemicals will store for a month or more and you will not need to weigh out the dyeing components each time you wish to dye your material.

The easiest concentration to use is a 2·5% solution of dyes and a 10% solution of chemicals.

PREPARATION OF A 2·5% DYE SOLUTION

Weigh out 5 g of dye powder and transfer to a pyrex flask or beaker. Add about 100 ml of hot water, stir in with a stirring rod to remove the lumps and boil for about 30 seconds. When cool, bring up the volume to 200 ml with cold water. Pour into the bottle and label as follows

| |
|---|
| Dye: ........................................................................................................<br>5 g in 200 ml<br>Date made up: ................................................................................... |

PREPARATION OF A 10% CHEMICAL SOLUTION

Weigh out 20 g of the required chemical, or if a liquid measure out 20 ml, add to the flask and bring up to a volume of 200 ml with cold water. Transfer to the storage bottle and label as follows

```
Chemical: ........................................................................
20 g in 200 ml
Date made up: ...............................................................
```

Formula for diluted dyes and chemicals is as follows

$$\text{Volume of solution required} = \frac{\text{Percentage}}{100} \times \frac{\text{Weight of Material}}{1} \times \frac{100}{\text{Solution strength}}$$

We can now look at an easy dyeing recipe and, assuming you have made up your solutions as shown, we can show the calculations necessary to prepare the dyebath for a 50 g batch of material. We will take the following recipe:

1·0% Acid Yellow 4 GLS*
0·5% Acid Red 3 BS
1·2% Acid Blue 3 RW
10% Sodium Sulphate
5% Acetic Acid
3% Leveller

Weight of wool: 50 g    Liquor Ratio 30:1

*Calculation using Standard Formula* (for diluted dyes and chemicals)

Vol. of Yellow 4GLS =
$$\frac{1 \cdot 0}{100} \times \frac{50}{1} \times \frac{100}{2 \cdot 5} = 20 \text{ ml of } 2 \cdot 5\% \text{ yellow solution}$$

Vol. of Red 3BS =
$$\frac{0 \cdot 5}{100} \times \frac{50}{1} \times \frac{100}{2 \cdot 5} = 10 \text{ ml of } 2 \cdot 5\% \text{ red solution}$$

Vol. of Blue 3RW =
$$\frac{1 \cdot 2}{100} \times \frac{50}{1} \times \frac{100}{2 \cdot 5} = 24 \text{ ml of } 2 \cdot 5\% \text{ blue solution}$$

Vol. of Sodium Sulphate =
$$\frac{10}{100} \times \frac{50}{1} \times \frac{100}{10} = 50 \text{ ml of } 10\% \text{ solution}$$

Vol. of Acetic Acid =
$$\frac{5}{100} \times \frac{50}{1} \times \frac{100}{10} = 25 \text{ ml of } 10\% \text{ solution}$$

Vol. of Leveller =
$$\frac{3}{100} \times \frac{50}{1} \times \frac{100}{10} = 15 \text{ ml of } 10\% \text{ solution}$$

Volume of dyebath at 30:1 and 50 g of wool = 1500 ml.
Using a pipette or other measuring device, you can measure into the dyebath the required amounts of dyes and chemicals calculated as above.

* For explanation of letter codes, see page 28.

### Concentration calculations

There will be some recipes that require the weight of chemicals to be calculated on the volume of the dyebath and not on the weight of material as discussed above. Here the recipe will be quoted as a *concentration* amount of grams per litre (g/litre). Before this step can be calculated we must first determine the bath volume from the liquor ratio calculation and then find how many grams are required under the recipe instructions.

The concept is simple. A 10 g/litre solution of a chemical contains 10 grams of that chemical dissolved in 1 litre of water, or the same proportions, eg. 5 grams of salt dissolved in 500 ml of water, would have a concentration of 10 g/1000 ml which is the same as saying 10 g/litre. Equally, if we had a solution of say, 5 grams dissolved in 200 ml, it would have an equivalent concentration of 25 grams in 1000 ml, which is 25 g/litre.

As an example, let us look at a simple formula for pickling sheep skins. The recipe calls for the following:

50 g/litre common salt
20 g/litre formic acid
L.R. = 30 : 1 Weight of skin 4000 g or 4 kilograms
Volume of pickle bath:  30 × 4 kg = 120 litres
Weight of common salt: 50 g/litre × 120 litres = 6000 g or 6 kg
Weight of formic acid:  20 g/litre × 120 litres = 2400 g or 2·4 kg

The rule then is to multiply the concentration in gram per litre by the bath volume and call the answer grams required.

### Acidity and alkalinity

Ignoring the physical chemistry associated with the concepts of acidity and alkalinity, you will need to know the scale, called the *pH scale*, that measures those properties.

Any chemical solution in water can be either acid, neutral or alkaline. Solutions of sulphuric acid are acid while solutions of ammonia or sodium carbonate are alkaline. Pure rain water or distilled water can be neutral as can a solution of sugar in water.

We can assign a scale of numbers such that very acid solutions are designated 1, very strong alkaline solutions designated 14, and neutral solutions fall in the middle at 7. Therefore the scale runs from 1 for the strong acid solutions, through 5 for the weak acids, to 7 for neutral through 9 for the weak alkalis, and to 14 for the strong alkalis.

Most dyebaths will be around the neutral point of between pH 6 and pH7 and where the recipe calls for you to check the pH it is easy to purchase pH indicating papers from the chemical supply houses. These are easily used by dipping the paper into the bath, waiting for the colour change to occur, and then comparing the colour of the paper to a reference colour on the packet which will indicate the acidity or alkalinity of the dyebath. (figure 1)

Fig. 1

**Dyeing vessel construction**
The dyeing operation requires heat to drive the dye from the solution to the material, and it is therefore necessary to have a dyeing container that will withstand heating to bring the bath to the boil. It also needs to be big enough to accommodate the size and weight of material to be dyed, and we have seen how one can calculate the volume needed dependent on the weight of textile. One other important factor in the specification of the dyepot is that it should be of non-reactive material. When you come to prepare the mordant baths you will see that both iron and copper can dull bright dyeings, and if these metals are allowed to come in contact with the dyeing operation, then we can expect a significant change in colour, due to the presence of metals. Therefore we would not dye in iron or copper pots.

Ideally the dyeing should be carried out in a Pyrex glass container, which can be obtained from chemical supply houses. Figure 2 shows the range of beakers and flasks needed if you were to dye 50 g of material using three dye solutions and three groups of chemicals.

You may have enamel basins around the house, and these will do as long as there are no chips in the enamel that would allow the metal to show through and disturb the dyeing process. Do not use aluminium as the chemicals used in the dyebaths will corrode the containers and cause lakes to be formed in the bath, influencing the colour. Stainless steel, if you can afford

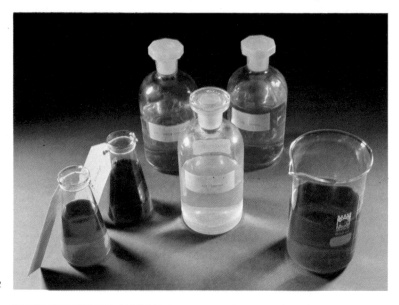

Fig. 2

Fig. 3

it, is one of the best materials for the dyebath, as it will not corrode, or shatter when dropped. Figure 3 shows a range of vessels constructed from stainless steel.

**Glassware**

The dyebath should be stirred as it is heated from cold to the boil, and if this is not done, uneven patches will appear, giving unlevel dye sections of the material. Therefore it is advisable to stir the bath, and the best material

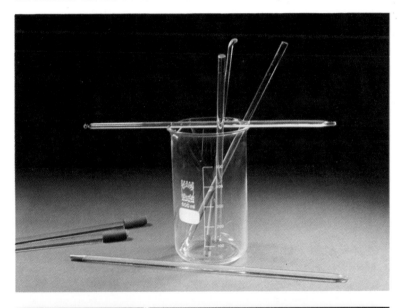

Fig. 4

Fig. 5

is glass rods tipped with some rubber tubing as shown in Figure 4. The advantage of glass over wooden dowel is that glass does not stain in strong dye solutions and does not splinter or become rough and snag fine materials like silk or rayon. The supply houses will provide these essentials.

Old coffee and sauce bottles should be kept for storing dye and chemical solutions. Label as shown in the previous examples.

Measuring equipment is important whether it be measuring cylinders, medicine dispensers or pipettes that can measure an accurate volume of solution to the bath. Figure 5 shows various measuring devices.

SAFETY WITH CHEMICALS

You will remember the television commercial where the point is made that 'oils is oils'. Well, 'chemicals is chemicals' as well, and should be respected as such. There are some chemicals in the dyers range that are not for human consumption, and will cause irritation if allowed to remain on the skin. There is an old saying that will bear repeating, as it has to every apprentice dyer the world over. 'An ounce of common sense per pound of yarn dyed'. It is good advice when you consider that the chemicals used in the various processes are dangerous if not treated with respect, and allowed to fall into very young hands.

Make it a habit to wash all containers with hot soapy water after use. Label all chemicals and dyes clearly, (figure 6) and keep some first aid chart at hand that will show you what action to take if the unforeseen happens. Keep the dyeing operation apart from the cooking operation if you are producing colours in the home, and do not use the same containers for cooking one day and dyeing the next.

CHEMICAL AND DYE SUPPLY SOURCES

You will no doubt wish to purchase dyes, chemicals and apparatus for your new hobby, and this is very easy to do,

At the end of this book there is a comprehensive list of supply houses throughout Australia and New Zealand. A word of caution when you approach dye companies. Suppliers are set up to supply the textile trade, and they are dealing with orders around an average of 50 kg to tonne lots per order. You should approach the company with this fact in mind and make quite sure you know how much you really want. For the home or school dyer, 50 g is adequate, providing you are not going to dye large quantities of material. If you ask for quantities around 50 g you will not cause them too much inconvenience. The Australian and New Zealand suppliers are very helpful, but remember they are doing you a favour outside their normal routine. The craft suppliers, on the other hand, are set up to deal with your problems and your supplies. They have dyes, chemicals, and the appropriate apparatus pre-packed and ready to be purchased off the shelf.

You should write to suppliers for a catalogue, sending a stamped self-addressed envelope, or if you know which products are required, order accordingly. It has been suggested that the craft dyeing operation is only a means to an end, and that weavers and spinners, knitters and macrame makers may not have the facilities, nor wish to outlay the money involved to purchase pyrex glass, stainless steel ware or expensive thermometers. I accept this argument but would make the point that the expense is usually 'one off', and having obtained the tools of trade, with careful handling, the various apparatus should last for many years.

There are alternatives offered as you will have seen from the section on glassware and dyeing vessel construction, and if these are more readily pro-

Fig. 6

cured then you should not concern yourself with the expensive apparatus. Similarly you may at this stage be a little frightened by the seemingly complex terms used and may not wish to become so involved in the process of dyeing.

You are urged to persevere and learn the art as you once learnt your weaving or spinning technology, and if you wish to translate difficult sections into more rudimentary basic formulae, then you should try this technique as well. If you obtain the desired result, then that is the important consideration. The book has been written for the advanced dyer as well as for the novice; the main consideration being to implant some science into the classical art form of craft dyeing.

# 2 Natural and man-made fibres

The fibres used in modern textile manufacture can be classified into two main groups: (a) those naturally occurring in nature, and, (b) those which are man-made. The *natural fibres* are those such as cotton, wool, silk and flax, and are provided by animals and plants in a ready-made fibrous form. The *man-made fibres* are those in which some manufacturing process has generated a fibre from either natural products or synthesised the fibres from the chemistry of carbon (organic chemistry).

**Natural fibres**
These fibres can be subdivided into three main classes, according to the nature of their source.

> Vegetable fibres
> Animal fibres
> Mineral fibres

*Vegetable fibres* include the most important of all fibres; cotton. Together with flax or linen, hemp, jute and other fibres. They have all been produced by the plant and built on the chemistry of cellulose, the building blocks of the plant world from which most structures are derived.

*Animal fibres* include wool, silk, and the speciality hair fibres such as vicuna, angora and alpaca. These fibres are produced by the living animal and are built on the chemistry of the protein molecule with the complex permutations from which the animal body is made.

The last group, *mineral fibres*, are of little importance to the textile trade, and the most useful member of the group is asbestos which is made into fire-resistant clothing and industrial products.

**Man-made fibres**
These fibres can be further subdivided into two classes:

> Regenerated fibres
> Synthetic fibres

*Regenerated fibres* are those in which the building blocks have been provided in some form other than fibrous, and by chemical manipulation have been made to form fibres. Trees provide an abundant form of cellulose which is, in itself not suitable for textile use as a fibre. Over the last century chemists have reacted the cellulose with a variety of chemicals and the fibre known as viscose rayon is probably the most useful fibre that has been derived from their research. Peanuts have been used to provide a protein extract used in the production of 'Ardil', a fibre not in use now. Seaweed has provided us with an immense source of algenic acid from which a fibre can be produced. Milk, soya beans, and maize are other naturally occurring products that have been modified to produce regenerated fibres.

*Synthetic fibres* are those in which the entire operation has been synthesised from organic basic chemicals. The raw materials are not naturally occurring and are often derived from by-products of the destructive distillation of coal or the cracking of petroleum. The first commercial synthesised fibre was nylon, and was produced by the DuPont Company in the United States after extensive research by a Harvard University professor, W. H. Carothers. The polyesters, known now as terylene and dacron followed soon after, then the polyacrylonitriles, known as orlon and dralon.

The relationship between the generally used fibres is shown in the following table:

**Natural**

| | | | |
|---|---|---|---|
| Vegetable origin | { | Bast fibres | Flax |
| | | Leaf fibres | Sisal |
| | | Seed fibres | Cotton |
| Animal origin | { | Hair fibres | Wool |
| | | Filament | Silk |
| Mineral origin | | Rock fibres | Asbestos |

**Man-made**

| | | | |
|---|---|---|---|
| Regenerated origin | { | Cellulose | Viscose Rayon |
| | | Cellulose ester | Acetate Rayon |
| | | Protein | Ardil |
| | | Misc. | Alginate Rayon |
| Synthetic origin | { | Polyamides | Nylon |
| | | Polyesters | Terylene |
| | | Polyolefins | |
| | | Polyurethanes | Spandex |
| | | Polyvinyl fibres:— | |
| | |   Polyacrylonitrile | Orlon |
| | |   Polyvinyl chloride | Saran |
| | |   Polytetrafluoro | Teflon |

## Fastness consideration in dyed textile materials

Ever since the introduction of the first man-made dye-stuffs over 120 years ago, there have been strivings after higher degrees of fastness of all kinds, both on the part of dyestuff manufacturers and dyestuff users. The desire of the textile trade to put on the market articles of 'guaranteed fastness' directed attention to the need for accepted standards against which fastness could be tested and assessed.

In the commercial world of dyestuff users and manufacturers, there are few, if any, dyes that have a perfect fastness to all the various exposures and treatments. Rather it is a question of 'horses for courses' and the selection of dyestuffs to suit a particular purpose is part of the dyer's skill. In the craft dyers field the choice of fastness is made by nature, and it is a sad fact that some of the most beautiful hues are transient and will not withstand exposure to light, heat or washing. There are exceptions, indigo being one of these, and cotton that has been dyed by indigo is very fast to almost any treatment given by the consumer. Madder as dyed by the traditional three week process had exceptional fastness.

For the home or school dyer, it is a simple process to determine the relative fastness to the dyed goods and a short summary is included for those wishing to evaluate their own processes. When using commercial dyestuffs the amateur dyer can predict the resulting fastness by consulting the manufacturer's shade card in which all the necessary information has been compiled from testing carried out in the dye manufacturer's laboratory. (figure 7)

## Light fastness testing

The professional dyer rates his dyestuff on a scale between one and eight. These gradings are arranged in ascending order of degrees of resistance to daylight fading. The home dyer can rate whether the dyed pattern will fade by exposing a sample to daylight, under glass and at an angle of 45° facing

Fig. 7

Fig. 8

Fig. 9

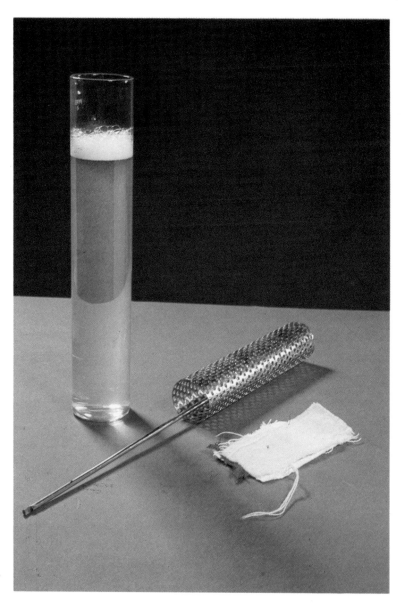

Fig. 10

due north. One quarter of the pattern is covered by cardboard so that it is shielded from the direct rays, the other three quarters being exposed. (figure 8) After each day's exposure, check the unexposed portion against the exposed, and assess the degree of contrast. You may find that after one day's exposure to average sunny conditions, the colour will have faded substantially. It will depend on the intensity of the sunlight, but you will be able to judge whether you have dyed your pattern fast or fugitive. Figure 9 shows a novel fastness test where half the pattern was covered by a stiff cardboard cover. The other half was covered by 5 separate transverse cardboard strips.

When exposure to sunlight commences, remove cover 1 and expose to sunlight for one day. Remove strip 2 and expose for a further one day. Remove strip 3 and expose for a further one day. Remove strip 4 and expose for a further one day. Remove strip 5 and expose for a further one day. Day one has little fading but, at day 5, the pattern has faded considerably. In their own right these fadings make an attractive and novel wall hanging.

**Washing fastness**

The extent to which any dye will lose colour or stain other adjacent fibres depends on the severity of the washing operation. Dyes that may not lose any colour when washed at 35°C may lose half of the depth or strength at a wash temperature of 65°C. The Australian Standards Association has set five standards tests, and the test method that is selected will reflect the end use of that particular dyed product. In other words one would not apply test 5 (colour fastness to very severe conditions of machine laundering), to a baby's delicate nylon layette. In this case the professional dyer would select Test 1 or maybe 2, colour fastness to hand laundering, or colour fastness to mild conditions of machine laundering. Again hospital linen goods would need to be dyed with dyes that could withstand Test 5.

For the home dyer, particularly when dyeing wool, test 1 or 2 is a satisfactory standard. In this test the goods are treated at either 40°C or 50°C with 5 g per litre of soap for either 30 minutes or in the case of test 2, for 45 minutes. The fibres to be tested are placed in an envelope made from cotton sheeting on one side, and woollen flannel on the other. A small piece of white nylon hand knitting yarn is placed in the envelope with the dyed sample as shown, and the top opening is stapled together. The test is placed in the soap solution and run at the appropriate temperature and time. After the test time, remove the sample bag from the liquor, rinse in cold water and dry in a warm place. You can assess the loss in strength of the washed sample by comparing it to the original untreated piece, and assess the degree of staining on the cotton side of the bag, and the degree of staining on the wool. Look at the nylon piece that you slipped inside. (figure 10)

If the dye is going to run on washing, chances are that the loose colour will stain the nylon first. If neither the wool side or the cotton side or the nylon is stained, you have successfully dyed your sample with a fast dye relative to that test. Do not worry if the soap solution is coloured, if it did not stain the wool, cotton or nylon, it will not stain any other fibre.

Fig. 11  **Rubbing fastness**
Depending on the type of dye applied and especially when dyeing craft vegetable extracted dyes, beware of a low rubbing fastness. You can test this easily by rubbing the dyed sample on a clean white cotton rag or handkerchief. Try the rubbing test wet and dry, and assess the staining on the white cotton. If it is bad, try treating the dyed sample in some warm soap solution, say about 2 grams per litre. There is a good chance that the loose dye will scour off, leaving a cleaner dyed sample. If the dye does not clean up, and still bleeds when rubbed, be careful and do not let the offending dyed material rub against light coloured goods. (figure 11)

# Scouring and Bleaching 3

This chapter describes the scouring and bleaching process for the main fibres covered in this book.

The efficient preparation of the material is a necessary phase in the dyeing operation, as the oils and greases present in the natural fibres will interfere with the application of the dyes and will cause poor fastness to rubbing and washing.

The scouring techniques apply more to the natural fibres with their relatively high level of impurities, and less to the synthetic fibres that only contain simple water-soluble spin finishes that can be removed by a low level scour. It is not always necessary to bleach these man-made fibres, as the produced colour on manufacture is a clean, white effect. The bleaching processes that are used to enhance this normal clean colour will not be discussed.

## Scouring Wool and Silk, Cotton and Flax, man-made fibres

### SCOURING OF WOOL

Natural fleece wools are heavily contaminated with a variety of substances, and a merino fleece will have up to 50% of its greasy weight attributable to products other than the wool fibre. We would expect on average about 20% of the greasy weight to be dirt, burrs and grass seeds. Perspiration products, commonly called 'suint', would make up 5%, there would be about 10 to 12% water, and the remaining 16% would be natural wool grease or lanolin.

If the wool is in the spun form it can be hanked before scouring between thumb and elbow, as shown in *Figure* 12. Tie two or three loose tie bands around the hank to reduce the danger of tangling.

The industrial wool scouring process utilises a four-bath process, and I have used that as a model for the craft dyer. In the following recipe, the liquor ratio is 10:1 and the quantities on the right side of the recipe are calculated for 1 kg of greasy wool.

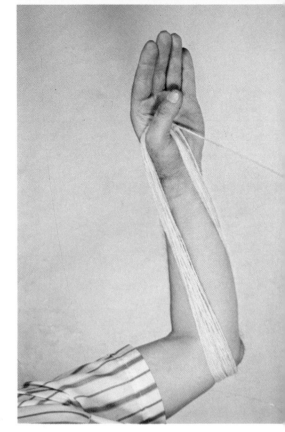

Fig. 12

Bath 1
    Pure soap powder    0·8%    8 g
    Sodium carbonate (Soda Ash)    0·2%    2 g or 20 ml of a 10% solution
    Water, to 10 litres. Temperature 50°C    Time of immersion: 3 minutes

Bath 2
    Pure soap powder    0·4%    4 g
    Sodium carbonate    0·1%    1 g or 10 ml of a 10% solution
    Water, to 10 litres, Temperature 45°C    Time of immersion: 2 minutes

Bath 3
    Pure soap powder    0·3%    3 g
    Sodium carbonate    0·1%    1 g or 10 ml of a 10% solution
    Water, to 10 litres. Temperature 45°C    Time of immersion: 2 minutes

Bath 4
    Water at 40°C for one minute.

Wool, either in loose form or in hanks, is worked gently in each bath, squeezing excess scour solution from the material before placing it in a new bath. The final grease content will be about 0·2% but the wool still contains burrs and grass seeds. If the wool is in the loose state, these can be carded out using the hand cards.

Dry the scoured wool on a towel away from direct sunlight.

SILK SCOURING AND DEGUMMING

Although silk is a protein fibre, it is treated in a different manner to wool. The scouring of silk is usually associated with removal of the silk gum, the natural glue that the silkworm has bound the twin fibres together with at the time of spinning. The goods are usually treated for about two hours in a 0·5% solution of soap at 95°C using a liquor ration of 30:1. Vigorous movement or exposure to rough surfaces, or excessive boiling, will cause the fibres to become ruptured and bundles of fibrils will project from the surface. These scatter instead of reflecting the incident light and give rise to patches differing in appearance from the undamaged silk. Once this form of damage has occurred it cannot be removed, and the silk is said to be 'lousy'.

COTTON AND FLAX SCOURING

Both fibres contain a fair proportion of natural wax and pectin, and a harsher treatment than was given wool must be used to remove these contaminants from the fibres. Cotton is scoured at the boil in the presence of caustic soda and detergent, which saponify the natural impurities and hold them in suspension. At the conclusion of a normal scour which composes of a caustic boil and bleach, the cotton has an improved water absorbency which will produce an even dye uptake. In the following example, the liquor ratio is set at 10:1 and the quantities on the right side of the recipe are calculated for 500 g of cotton or flax.

Scouring Recipe (can be repeated twice if necessary).

| | | |
|---|---|---|
| Caustic Soda | 1·5% | 7·5 g or 75 ml of a 10% solution |
| Detergent type A | 1·5% | 7·5 ml |

Water to 5 litres, temperature at the boil. Immersion time 2 hours. Top up water as evaporation lowers the level.

At the conclusion of this treatment, rinse the cotton under hot running water for about 15 minutes and then for 15 minutes in cold water. Acidify the cotton in a fresh bath, same volume, using acetic acid. Adjust the acidity or the pH of the bath to 6·0. This will take about 50 ml of acetic acid.

At this stage the cotton will possess a cream to fawn coloration which can be bleached, and you should refer to the later pages of this chapter to follow the bleaching process.

MAN-MADE FIBRE SCOURING
The impurities, as we have seen, are only spinning finishes and collected dirt. They can be removed by a simple detergent scour consisting of 1·5% detergent together with 1·0% sodium carbonate for about 30 minutes at about 50–60°C. With all wet treatments of synthetic fibres at temperatures in excess of 40°C, rapid cooling will crease and heat-set the material. At the conclusion of any temperature treatment, cool the bath *slowly* with gentle stirring, being careful not to heat-set the creases.

**Bleaching**
*Wool and silk*
Depending on the type of fibre and the origins of its formation, the two fibres show, within their various classes, a wide variety of natural coloration. The base silk colour varies from lot to lot, depending on the growing conditions. With wool, the colour can range from an off white for scoured merino wool, to a dirty grey for the coarser crossbred qualities. The natural base colour is usually satisfactory if you wish only medium to dark shades. If, however, you wish to dye the paler tones, and when using tin mordants on some of the natural dyes, a bleached base gives a cleaner appearance to the dyed wool.

The simplest and most widely used method consists of an overnight immersion in a solution of hydrogen peroxide and alkali. You must keep the fibres underneath the surface of the bath at all times and this can be achieved by weighting the hanks with china cups or cereal bowls. Do not use metal bleach baths, apart from stainless steel, as the hydrogen peroxide will produce oxygen at a faster rate than intended, and the bath will be exhausted and flattened before the bleach has had a chance to oxidise the wool or silk colouring.

*Care* must be taken with hydrogen peroxide; it is a very strong oxidising agent and if spillage does occur, wipe up immediately with a wet rag. Wear gloves and protect the eyes from splashes. Keep away from small children and label the container accordingly.

The following recipe is a two-bath process, and is calculated on 100 g of wool or silk with a liquor ratio of 30:1.
Prepare the following chemicals in the bleach bath:

| | |
|---|---|
| Hydrogen peroxide (35% solution strength) | 18 ml |
| Tri sodium pyrophosphate | 2 g |
| Ammonia (household type cloudy ammonia) | 6 ml |

Set the temperature to about 50°C and weigh down the hanks under the liquor level. Leave overnight, about 12 hours.
Next morning, wash off with cold water and prepare the second bath as follows:

| | |
|---|---|
| Sodium metabisulphite | 3 g |
| Detergent type B (Softly®) | 2 ml |
| Calgon® powder | 1 g |

Set the bath at 50°C for two hours. At the conclusion of the second bath, rinse the material in warm running water, dry on a towel away from direct sunlight.

*Cotton and Flax*

The bleaching of cotton and flax can be carried out with either hydrogen peroxide or sodium hypochlorite, or a combination of both. Sodium hypochlorite is sold under the commercial name of White King® and is a solution of 4% sodium hypochlorite. The recipe as given will use this diluted form of the hypochlorite.

The following recipe is a two-bath process, and is calculated for 500 g of cotton or flax with a liquor ration of 30:1.

Prepare the following chemicals in the bleach bath:

| | |
|---|---|
| Sodium hypochlorite as White King® | 19 ml |
| Sodium carbonate (soda ash) | 120 g |

The temperature is cold and the immersion time is 30 to 45 minutes. At the conclusion of the bleaching process, rinse under a cold tap for 5 minutes.

The second bath is prepared as follows:

| | |
|---|---|
| Hydrogen peroxide (50%) | 150 ml |
| Caustic soda | 5 g |
| | or 50 ml of a 10% solution |
| Sodium silicate | 60 g |

The temperature is raised to 90°C over 45 minutes and held at 90°C for 60 minutes. Rinse under warm water for 5 minutes.

Softly˟ registered trade mark of Lever & Kitchen
Calgon˟ registered trade mark of Albright & Wilson
White King˟ registered trade mark of Kiwi Australia Ltd.

Chemicals used in the scouring and bleaching process as described in Chapter 3

| Name | Commercial Product | Availability | Hazard |
|---|---|---|---|
| Ammonia | Cloudy Ammonia | Supermarkets | †† |
|  | Ammonia solution | Chemists | ††† |
| Sodium carbonate (Soda ash) | Washing soda | Supermarkets |  |
| Detergents type A | Comprox® | Service stations |  |
|  | Trix® | Supermarkets |  |
| Detergents type B | Softly® | Supermarkets |  |
| Caustic soda | Caustic soda | Chemical supply | ††† |
| Acetic acid | Acetic acid | Chemical supply | †† |
|  | White vinegar | Supermarkets |  |
| Sodium hypochlorite | White King® | Supermarkets | †† |
| Hydrogen peroxide | Hydrogen peroxide (35%) | Chemists | ††† |
| Tri sodium pyrophosphate | Tri sodium pyrophosphate | Chemical supply | † |
| Calgon® | Calgon® | Supermarkets |  |
| Sodium silicate | Water glass (Egg preserving compounds) | Supermarkets or chemists | † |

Hazard Code:

††† **HANDLE WITH CARE** — Corrosive to skin. Wash off with running water if spilt. Keep away from eyes and do not inhale fumes. Do not ingest. Call doctor if swallowed. Neutralise with milk and bicarbonate of soda.

†† **REASONABLE CARE** — Wash off from skin if spilt. Do not inhale fumes.

† **CARE** — Wash of from skin if prolonged contact.

Comprox® registered trademark of B P Australia
Trix® registered trademark of Reckitt & Colman Pty Ltd.

# Methods of application for commercial dyes  4

Most textile fibres are comparatively easy to dye and obtain bright fast shades. It is complicated by the restrictions of levelness and recipe continuity, fastness and the multitude of various fibres, dyed either by themselves or in mixtures. These considerations, which are vital to the manufacturing dyer, need not concern us in this book.

With most fibres, the dyeing operation is a matter of adding dyestuff powder to fresh water, adding some chemicals and the material to be dyed, and then heating the bath to the boil and leaving it at this temperature for about 45 minutes. As you will not have the advantage of reference to the manufacturer's pattern cards in which he shows you an actual dyeing at different depths and all the fastness details and the variations of methods, you will have to rely on the following formulations and 'blind dye' samples until you have gained some proficiency in the dyeing operation. At the start you will have no guide to the actual shade produced on each fibre until you complete the whole operation, and so it will be rather like dyeing with natural dyes where the colour produced depends on the time of the year that you picked the flowers or weeds used in the dyeing process. Should you wish to pursue dyeing as a hobby, you can approach the dye manufacturers and ask for their pattern cards and formulate the exact colours you wish to copy and then learn to predict the formulations necessary. (figure 13)

In this chapter the messy and complicated dye procedures are ignored and the technically sophisticated side of the operation that is a part of life in the commercial dyehouses has been omitted. You need to do only two technical operations.
1 Make up your dye powders in a water solution, and
2 Make up the chemicals and auxiliary reagents also in solution form. If you can follow simple cooking instructions successfully, then you will become expert in simple dyeing very quickly.

DYESTUFFS
The powders as sold by the manufacturer are dusty, they stain fibres and skin, and are not easy to work with. It is a far better idea to prepare a known concentration in water and use the required amounts for your dyeing operations. If you make up three different solutions, a red, yellow and a blue,

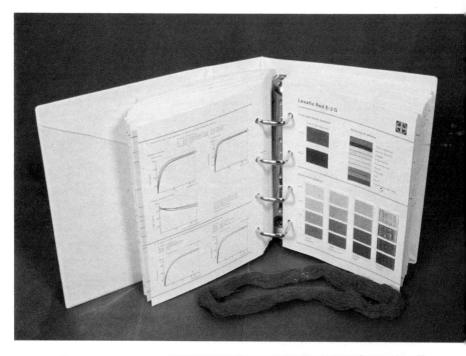

Fig. 13

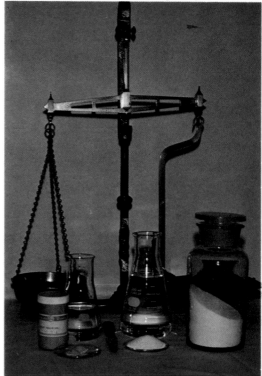

Fig. 14

you can combine different quantities to obtain just about any colour on the spectrum. For a violet, mix blue and red, similarly for an orange, yellow and red, etc. It's that simple.

Assume that you have obtained some dye powder, it does not matter what it is, and that you have about 50 grams of material that you want to dye. You will by this stage have clearly labelled the small bottle of dye powder similar to the label as shown below:

> Acid dye Red 3 FGA

Weigh out about 5 g of dye powder and transfer to a Pyrex flask or beaker. Add 200 ml of hot water and stir to dissolve the lumps, then boil for about 30 seconds and place the dye solution to one side to cool. The strength of this solution you have just made is 2·5%, i.e. 5 g in 200 ml. Do this for all three colours so that you have three flasks with different dye solutions. Label the flasks with luggage labels around the neck. (*figure 14*)

CHEMICALS AND ASSISTANTS

The dyeing operation also requires acids, salts like ammonium sulphate and sodium sulphate, as well as different proprietary dyeing assistants necessary to transfer the dyestuff from the water solution that makes up the dyebath, on to the fibre. As with the dyes, make up these reagents in solution, but this time make up a 10% solution.

Weigh up 10 g of reagent and dissolve in 100 ml of water, or if you are going to carry out a number of dyeings, make up 100 g in 1 litre. Mix the reagents with warm water, say about 40°C and then label the bottle as such:

> 10% Sodium Sulphate Solution

THE DYEING OPERATION

Unless otherwise stated, the basic dyeing operation is as follows. (Where differences do exist, the text will indicate how and why.)

1  Calculate the liquor ratio from the amount of material you wish to dye.
2  Add this amount of water to the dye vessel.
3  Add the chemicals and assistants as calculated.
4  Add the material to be dyed.
5  Add the dyestuff solution as calculated.
6  Raise to the boil over 45 minutes, gently stirring.
7  Keep at the boil, just simmering, for about 45 minutes.

Nothing too difficult in that is there? Let us now look at the method of calculation and some examples. Refer back to Chapter 1 and revise the basic concepts once more.

## LIQUOR RATIO AND BATH VOLUME

Multiply the dry weight of yarn or fabric by 30. This will give you the number of ml needed for the bath volume. For 50 g of material — 50 g × 30 = 1500 ml of cold water.

## DEPTH OF SHADE

For pale shades use about 0·2% of dye and calculate as shown:

$$\frac{0\cdot 2}{100} \times 50 \times \frac{200}{5} = 4 \text{ ml of your } 2\cdot 5\% \text{ dye solution}$$

For average to medium depths of shade use about 1% or 20 ml, and for heavy shades about 4% to 6% which would be 80 to 120 ml.

## AMOUNTS OF CHEMICALS AND LEVELLING AGENTS

As we discuss the various types of dye classes, reference will be made to the amounts of chemicals and auxiliary assistants to be used. The text will indicate the percentage of each reagent to be used.

*Example:* 3% Ammonium Sulphate

Take your previously made 10% solution of ammonium sulphate and measure out 15 ml. The calculation was the same as that for the dye solution.

$$\frac{3}{100} \times 50 \times \frac{100}{10} = 15 \text{ ml Ammonium Sulphate (10\% solution)}$$

That is the most difficult calculation to make in the whole book. The rest of the text contains descriptions of the types of dye classes and their application methods to the various fibres.

Let us now look in detail at some of the dye classes that are used in the coloration of textile fibres. Dye manufacturers usually use a coded description to identify each dye produced. There is no secret to this code and it allows one to compare dyes without actually dyeing them out.

Take for example the following description:

Supranol Fast Yellow 4R 200

Reference to the Table on page 30 will indicate that this is an acid milling dye produced by Bayer.

*Supranol Fast* tells us that it is a particular class of acid milling dyestuff.

*Yellow* describes the hue of the dyestuff.

*4R* indicates that it is quite red, redder than say Yellow 2R. Here the R stands for *Rot* which is the German translation for Red. You will sometimes see identifiers using a G which stands for *Gelb* or yellow; B for *Blau* or blue. Some codes used in this section of the name do not refer to the tone of the dye but to some distinguishing fact applicable to that dyestuff.

The last part of the code is the figure indicating the comparative strength of the dye. The manufacturer of the dye powder formulates his batch to a very strong strength. The dye as produced may be so strong that it could have serious limitations in the actual dyeing process. It may be so strong that it will not dissolve readily in water, it may also have a tendency to dye unlevel, and so the manufacturer will then dilute his dye powder with dilutents such as dextrin, salt, dispersing agents, and adjust his production batch to a predetermined known dye strength. Let's assume he dilutes the powder down to a recommended dilution powder form and sells it to you at what he calls 'standard strength'. He will most likely place an 'S' on the end of the identifying code such that the description will be

Supranol Fast Yellow   4RS

He may decide to make a stronger version, say twice as strong. You, the dyer, would use half as much dye of this stronger powder than you would have used if you were formulating with the 4RS quality.

To differentiate between the strong and weak version he would then call the stronger dye

Supranol Fast Yellow   4R 200

If he made a version six times as strong he would identify this batch as

Supranol Fast Yellow   4R 600

**Wool and related fibres**
There are five main dye classes applicable to wool, most can be used for dyeing silk and all can be used to dye nylon. So with wool colours you get a bonus.

> The dye classes are
> Acid Levelling dyes
> Acid Milling dyes
> Premetallised dyes
> Chrome mordant dyes
> Reactive dyes.

*Acid Levelling* dyes are probably the most simple of dyestuffs to apply. They are bright colours but are not very fast to washing or light. There are exceptions, some will withstand strong sunlight for a reasonable time, others are non-fast to moonlight!

*Acid Milling* dyes are faster to washing and light, and form the main basis, with premetallised colours, for the commercial dyeing of woollen goods. They are harder to dye level than the level acid dyes, but with care and adequate stirring they will dye even. They are usually bright shades.

*Premetallised* dyes are rather dull but are very fast to wet treatment and to light. They are relatively easy to apply to wool, silk and nylon, but harder to rectify if the colour is uneven after the first dyeing.

*Chrome Mordant* dyes are similar to dyeing with natural dyes except that the mordant can be applied either before or after the normal dyeing operation. Like natural dyes, the Chrome Mordants require chromium salts to develop the colour and to impart a high degree of wash fastness. They are cheap but demand a lengthy process time.

Reactive wool dyes are perhaps the ultimate dyeing class for wool, and chemically combine with the wool molecule to give exceptionally fast dyes. They are just about impossible to level if they go wrong, and are the basis for the shades used in the Superwash wool range.

Listed below are some trade names that are used by the dyestuff manufacturers to distinguish between the various dye classes. The list is of course not exhaustive and not all manufacturers are shown.

|  | Bayer | Sandoz | ICI | Ciba-Geigy |
|---|---|---|---|---|
| Acid Levelling | Supracen | Sandolan E | Lissamine | Erio |
| Acid Milling | Supranol | Sandolan | Coomassie | Polar |
| Premetallised | Isolan | Lanasyn | – | Irgalan |
| Chrome Mordant | Diamond | Omega | Solo-chrome | Erio-chrome |
| Reactive | Verofix | Drimalan | Procilan | Lanasol |

An acid levelling dye from ICI would have Lissamine as the prefix followed by the dye description, e.g. Lissamine Scarlet 4 BS, whilst a Sandoz Chrome Mordant yellow could be described as Omega Chrome Flavine.

CHEMICALS NEEDED TO DYE WOOL, SILK OR NYLON

*Sodium Sulphate* can be purchased readily in Australia and is very cheap. A 10% solution is made simply by dissolving 10 g in 100 ml of cold water. The salt is non-poisonous and there are no precautions for its use.

*Acetic Acid.* Also cheap and plentiful. Can be purchased either as a concentrated 95% solution or a 60% solution. For the purposes of these exercises, if you use the 95% acid, dissolve 10 ml in 100 ml of cold water. Do the same if you can only get the 60% acid, but use one third more acetic acid in the dyebath. As an alternative, you can use white vinegar, but use about 10 times more than your calculations suggest.

**Caution** Concentrated acetic acid will burn if left on the skin, but can be easily washed away with water.

*Ammonium Sulphate.* Similar to sodium sulphate, 10 g in 100 ml of cold water. Ammonium sulphate is also safe and will not cause any burns. (It is good for lawns, and makes them nice and green in the hot summer months if well watered.)

*Levelling Agents.* Each manufacturer makes his own levelling agents and advises on their use. If you are dyeing with the Acid milling or the Premetallised dyes and you want a perfectly even and non-patchy result, then levelling agents are a must. They are complex chemicals built rather like a detergent and usually non-toxic. When you order your dyestuffs, ask the representative concerned if he will advise you on the one needed for the particular dye class. Albegal W for Acid Milling dyes, and Albegal SW for Premetallised dyes are recommended by Ciba-Geigy, Sandoz recommend the Lyogen range, and ICI, the Dispersol range of levellers. A 10% solution is made the same method as the chemicals, 10 ml in 100 ml of water and stir to dissolve.

*Formic Acid* is an acid that has to be handled with care, as it will burn skin if not washed off with cold water. It is considered a toxic chemical in its concentrated form. Diluted it is more docile, but you should handle with care. It is not a chemical to be frightened of, just respect it and you will have no trouble. It usually comes as a 95% strength and can be purchased from supply shops. Dilute it to a 10% solution by measuring out 10 ml of the concentrated acid and dilute to 100 ml.

**Recipes for dyeing wool, silk and nylon**

All chemical amounts calculated from a 10% solution.

ACID LEVELLING DYES
Recipe for 50 g of material    L.R. = 30:1

| Add to the cold bath, | Sodium sulphate | 20% | (100 ml) |
| --- | --- | --- | --- |
| | Formic acid | 4% | (20 ml) |

pH should be about 3-4 (check with papers).
Add the material and stir for about 3 minutes, then add the required amount of dye solution and begin to heat the bath. Stir gradually as you bring the dye-bath to the boil over 30 minutes. Let boil for a further 30 minutes then rinse the material and dry.

ACID MILLING DYES
Recipe for 50 g of material    L.R. = 30:1

| Add to the cold bath, | Sodium sulphate | 10% | (50 ml) |
| --- | --- | --- | --- |
| | Acetic acid | 3% | (15 ml) |
| | Levelling agent | 2% | (10 ml) |

pH should be about 5·5-6·5 (check with papers).
Carry out the same procedure as in Acid Levelling dyes but boil for about 45 minutes before rinsing.

PREMETALLISED DYES
Recipe for 50 g of material    L.R. = 30:1

| Add to the cold bath, | Ammonium sulphate | 10% | (50 ml) |
| --- | --- | --- | --- |
| | Levelling agent | 2% | (10 ml) |
| | Acetic acid | 3% | (15 ml) |

Carry out the same dyeing procedure as above, boiling for 45 minutes.
pH should be about 6-7

Chrome dyes and Reactives are complex to dye on wool and do require extra processing compared to the previous methods. Both types of dyes are used extensively on wool, nylon and to some extent on silk. Their exceedingly good wash-fastness and light-fastness make them useful to most sections of the trade.

CHROME DYES

After chrome process. (figure 15)

Recipe for 50 g of material     L.R. = 30:1

Add to the cold bath:  Sodium sulphate     10%   (50 ml)
                       Acetic acid          3%    (15 ml)
                       Levelling agent      2%    (10 ml)
                       Dye solution

Raise to the boil over 45 minutes and boil for 30 minutes. Add to the boiling bath, sulphuric or formic acid 1% (5 ml) and continue boiling until the bath is clear and free from colour, (usually about 20–30 minutes). Cool to about 80°C.

Add to the dye bath sodium dichromate 0·5% (2·5 ml) and continue boiling for another 30 minutes before rinsing.

REACTIVE DYES

Recipe for 50 g of material     L.R. = 30:1

Add to the cold bath:  Sodium sulphate     10%   (50 ml)
                       Acetic acid          3%    (15 ml)
                       Levelling agent      3%    (15 ml)
                       Dye solution

Raise to the boil over 45 minutes and boil for a further 30 minutes. If the depth of shade is over 1%, cool down to 80°C and add ammonia to bring

Fig. 15

the bath pH to 8·5. Keep at this temperature for a further 15 minutes and then cool gradually.
Rinse in warm water and in a fresh bath add:

Acetic acid 5%     (25 ml)

Allow wool to stay in the acetic acid bath for 15 minutes, remove from bath and rinse well. If the depth of shade was under 1%, wash off at the conclusion of the dyeing operation.

*Extra notes on Nylon and Silk*: Nylon will always dye deeper than the same shade depth on wool and so it will be necessary to decrease the proportions of dye (not chemicals or auxiliaries) if you wish to match the same shade on nylon that you had dyed on the wool.

Silk dyeings require added care to that given to wool or nylon, and apart from the more delicate nature of the fibre, on average the dyes tend to be less fast than the same shade on wool. Do not exceed a temperature of 85°C and the duration of time at this temperature should be restricted to 20 minutes. Silk fibres can be protected from delustering by the addition of soap solution or the silk scouring solution (see chapter 3) to the dyebath. 10% of the dyebath volume should in fact be this scour or 'boil-off liquor' as it is called. Apart from those restrictions, silk can be dyed in a similar way to wool or nylon, using the same recipes.

Cotton dyes, called Direct Dyes, will also dye silk if the boil-off liquor is present.

DIRECT DYE ON SILK
Recipe for 5 g of material     L.R. = 30:1
Add to the cold bath (containing at least 10% of its volume, the boil-off liquor):

Sodium sulphate     30%     (15 ml)
Soap solution        1%      (0·5 ml)

Add the material and stir for about 10 minutes, add the required amount of dye solution and raise to 85°C over 45 minutes. Keep at this temperature for 20 minutes.

**Cotton and linen fibres**

There are six main dye classes applicable to cotton and linen. The dye classes are as follows:

Direct dyes
Azoic dyes
Reactive dyes
Vat dyes
Soluble vat dyes
Sulphur dyes

*Direct dyes* are akin to the acid levelling class in wool. The directs are inexpensive and easy to apply, and although of indifferent wash-fastness, their use spread with great rapidity in the early days of synthesised dyestuffs. They are still used in great quantities. The dyes are soluble in water and can be applied to every form of cotton by simply boiling with the addition of common table salt, or sodium sulphate. These dyes will also dye silk.

*Azoic* colours are formed in the fibre itself by impregnating with an organic chemical called a naphthol, and then coupling with another complex chemical called a diazotised amine. There is no need to explain the chemistry of these fascinating dyes nor describe the complex dyeing cycle. Just watching an Azoic dye being applied to a cotton fibre is a joy to behold.

*Reactive dyes.* These are the latest class of dye to be added to the cotton range. Discovered in 1956, the ease of application coupled with the extremely high fastness to light and wet treatments make them a very important dye family, and they revolutionised the dyeing industry.

*Vats.* A complex dyeing system. One starts off with an insoluble powder, treats the powder with caustic soda and sodium hydrosulphite and the dye becomes soluble and will dye the cotton. Take the cotton out of the dye-bath and expose it to the air for a few minutes while the dye changes shade, and converts into the insoluble form inside the fibre giving an extremely fast dyestuff that will withstand very harsh wet treatments and bleaches.

*Soluble Vat* dyes are not used so much these days because of the price, but in their time were responsible for the pale shirting colours. Their use has been taken over by the reactive dyes.

*Sulphur Dyes* are rather like the Vat dyes, a little messier and the process tends to smell, due to the fact that sodium sulphide is used. People living next to sulphur dye works do have a lot to contend with!

The following table will outline the dyestuff manufacturers' trade names of the above dye classes that are applicable.

|  | Bayer | Sandoz | ICI | Ciba-Geigy |
|---|---|---|---|---|
| Direct | Sirius | Solar | Chlorozol Durazol | Solophenyl |
| Reactive | Levafix | Drimarine | Procion | Cibacron |
| Vat | Indanthren | Sandothrene | Caledon Durindone | Cibanone |
| Soluble Vats | – | Indigosol | Soledon | – |
| Sulphur Dyes | – | Thional | Thionol | – |

CHEMICALS YOU WILL NEED TO DYE COTTON AND LINEN
*Sodium sulphate* as discussed in the previous section.
*Sodium chloride*: Make up a 10% solution as for sodium sulphate.

*Sodium carbonate*: Sometimes called soda ash or washing soda. A 10% solution as for the previous two chemicals.

*Sodium hydrosulphite*: Not a pleasant chemical to work with due to its smell, but if you are contemplating dyeing Indigo or the Vat colours, then it is most necessary.

*Sodium hydroxide*: Commonly called caustic soda. Care must be taken when handling this alkali, as it will burn skin. When dissolving in water take extra care as the heat generated will boil the water if too much is added too quickly. Toxic, but reasonably safe in a 10% solution.

All chemicals can be prepared as a 10% solution by weighing out 10 g and dissolving in 100 ml of cold water.

RECIPES FOR DYEING COTTON OR LINEN

*Direct type dyes*
Recipe for 50 g of material     L.R. = 30:1
Add to the cold bath —  Common Salt 20%     100 ml
                        Dye solution

Add the cotton, stir well while raising the temperature to 85–90°C in about 15 minutes, and maintain this temperature for 60 minutes. Then remove the cotton, rinse well in cold water and dry.

*Reactive type dyes*
In this recipe the chemicals are added as the dyeing proceeds, and the amounts of chemicals are calculated on the volume of liquid and not on the weight of material. Weigh the chemicals AS POWDER and add them when instructed.

Recipe for 50 g of material     L.R. = 30:1    1500 ml
Chemicals needed:     Common salt           50 g/litre     75 g
                      Sodium carbonate      5 g/litre      7·5 g

Prepare the dyebath with the required dye solution and add the material to the COLD bath. During the first 30 minutes of dyeing add the common salt gradually in small dissolved portions. Then after a further 10 minutes add the sodium carbonate solution all at once and continue dyeing for a further 45 minutes. Finally rinse the material in cold water, boil in a fresh bath containing 3 g/litre of detergent such as Lissapol NC (ICI), rinse again and dry.

*Vat type dyes for cotton*
Because the vat dyes are a little more complex in their dyeing method, the process will be explained in detail. Also in this section the application of Indigo will be described.

*Method for Vat dyes* (warm dyeing types) for 5 g of material
Weigh out the 5 g of the selected vat dye, paste in a beaker with a little warm water and add 100 ml of cold water. For a 4% (medium to dark shade) dyeing on 5 g of cotton, place 20 ml of the dye solution in a test tube with

5 ml of the 10% sodium hydroxide solution and 1 ml of the sodium hydrosulphite solution. Warm the mixture to 60°C in a water bath, or place the test tube inside a larger container with water at a temperature of 60°C. Maintain at this temperature for 10 minutes. The solution will become clear and usually a different colour from the original 100 ml made up.

Next, prepare the dyebath at a L.R. of 50:1 or, 250 ml of water at 60°C and use this as a water bath to keep the test tubes at an even temperature during reduction. After 10 minutes, tip the contents of the test tube into the 250 ml dyebath, to which has been added 4 ml of caustic soda, and 4 ml of the sodium hydrosulphite solution. This removes any oxygen from the dyebath which would produce uneven results. Enter the wet cotton into the warm dyebath, maintaining at 60°C for 45 minutes. The patterns should not be allowed to protude above the liquor level and should be stirred gradually throughout the dyeing process. Finally, take them out of the liquor, squeeze without rinsing, hang in the air for 10 minutes to oxidise fully, then rinse in dilute acetic acid bath to neutralise the remaining alkali, rinse in hot water at 60°C and then boil in a soap solution (2 g/litre) for 15 minutes, rinse and dry.

The cold dyeing vats are similar, but the dyeing operation is carried out at 25°C, not 60°C as in the warm types. Also you must use 20% common salt as an assistant. Add the salt gradually, starting after the dyeing has progressed for 20 minutes, and throughout the processing time.

*Indigo on Cotton*
Indigo is one of the oldest of all known dyes and is obtained, from the plant *Indigofera tinctoria* grown in Asia and India. The synthetic product is now used for dyeing denim. Again the recipe is for 5 g and can be calculated for any weight of cotton that you wish to dye. Simply scale up each quantity.

Set the bath with 500 ml of cold water, 6 ml of the 10% sodium hydroxide solution and 1.5 ml of the sodium hydrosulphite. Add 2.5 g of Indigo grains (colour index number Vat Blue 1), stir gently till dissolved and allow to stand for 15 minutes, immerse the yarn and dye cold for 10 minutes. Remove and squeeze evenly by hand and expose to the air for 5 minutes. Return the pattern to the bath, dye for a further 10 minutes, squeeze well, and air oxidise for 20 minutes. Rinse in cold water and dry.

For a stronger dyeing use 20 ml of water, 3 ml of caustic solution, 1·5 ml of sodium hydrosulphite and 2·5 g of Indigo powder, stir as in previous example and stand for 15 minutes. Make up to 200 ml with cold water, allow to stand for 30 minutes. Immerse the cotton and give alternate dips and air oxidation as follows, (squeezing before each oxidation period):

**1st dip** 20 minutes immersion followed by 10 minutes oxidation
**2nd dip** 2 minutes immersion followed by 5 minutes oxidation

**3rd dip** 30 seconds immersion followed by 20 minutes oxidation
**4th dip** 30 seconds immersion followed by 5 minutes oxidation.
Dry without rinsing.

Vat dyes and indigo have been used for fast light- and very fast washing-fastness. As you have seen, they are insoluble in water, but by reduction with sodium hydrosulphite in the presence of caustic soda, they are converted into a soluble form which will dye cellulose. After dyeing, the air oxidation converts the soluble colour into an extremely fast colour.

**Viscose rayon**
RECIPES FOR DYEING VISCOSE RAYON
Viscose rayon is a regenerated cellulose fibre discovered in 1892, and today is still a major fibre on the world market. Being cellulose it is similar to cotton or flax and can be treated in a similar manner. There are two main methods discussed in this section.
DIRECT DYES ON VISCOSE RAYON
As for cotton with direct dyes (see previous section).
VAT DYES ON VISCOSE RAYON
As for cotton with vat dyes (see previous section).
DISPERSE DYES AND DYEING CELLULOSE ACETATE AND NYLON
The disperse dyes arose out of the endeavours of many workers to find an easy and commercially satisfactory way to dye cellulose acetate. Cellulose acetate was the first hydrophobic (water-hating) man-made fibre and is developed like viscose rayon from cellulose. Today disperse dyes are used for dyeing polyesters, e.g. terylene, dacron, and while these fibres will not be discussed, some knowledge of the disperse dyes, which can also be applied to nylon and orlon, should be included at this point.
Unlike all the other dyestuffs discussed so far, disperse dyes are *insoluble* in water. Therefore they are not dissolved in water but dispersed as fine particles throughout the solution, hence the name disperse dyes.
DISPERSE TYPE DYES ON CELLULOSE ACETATE
Recipe for 50 g of material    L.R. = 30:1
Add to the cold bath:
          Soap              1 g/litre
          Dye solution
Add the cellulose acetate, stir well while raising the temperature to between 70 and 85°C. Dyeing is continued at this temperature for 60 minutes or until the shade is level.
Light fastness about 5 - 6.
Reasonably fast to a mild wash.

DISPERSE TYPE DYES ON NYLON
As for cellulose acetate.

The light fastness of most disperse dyes on nylon is within the range of 4 to 6, although there are quite a few falling lower. The washing fastness varies considerably but can be as low as Grade 2 to a mild wash. They should not be used for heavy shades.

DYE SELECTION FOR CELLULOSE ACETATE AND NYLON

| | |
|---|---|
| Dispersol or Duranol range | I.C.I. |
| Cibacet | Ciba-Geigy |
| Artisil | Sandoz |
| Setacyl | Yorkshire Dye Company |

Preparation of your own pattern card is best carried out in the early stages of dyeing. You can explore the shade differences of each new dye purchased and experiment on the colours produced with other members of its class. For instance, you may have purchased some acid levelling dyes and wish to determine the shade ranges that can be produced from the combinations and permutations of these dyes. Pick a yellow, red and blue dye powder, dissolve them into standard solutions and prepare dyeings as shown on the colour triangle, Figure 16, which is built on a total of 1·0% colour depth. Look at any square and you will see three co-ordinates which are the percentages of 3 colours. For example, the fifth row, fourth square from the left indicates

$$Y = 0·2$$
$$R = 0·2$$
$$B = 0·6$$

which translated to a dyer's recipe, based on a 1% depth becomes

Yellow = 0·2%
Red = 0·2%
Blue = 0·6%

You will be able to determine from the total shades dyed, close approximations to the colour you desire to reproduce.

By proportioning the above formulas you can produce deeper or lighter triangles.

Figure 17 shows the trichromatic pattern card based on three dyestuffs, yellow R, scarlet G and blue G, and is a triangle where each dyeing is shown at a 2% depth.

Such triangles should be kept for reference to determine the possibilities of each trichromatic (three-colour) mixture. Well, there it is. A good sample of dyeing methods that should keep the average home dyer content for months. The dyeing process as stated in the opening paragraphs is a simple process. The preparation of the chemicals is time consuming, but once done and a large batch made up, it is not so difficult the next time.

Yellow

Y = 1·0
R = 0·0
B = 0·0

Y = 0·8     Y = 0·8
R = 0·2     R = 0·0
B = 0·0     B = 0·2

Y = 0·6     Y = 0·6     Y = 0·6
R = 0·4     R = 0·2     R = 0·0
B = 0·0     B = 0·2     B = 0·4

Y = 0·4     Y = 0·4     Y = 0·4     Y = 0·4
R = 0·6     R = 0·4     R = 0·2     R = 0·0
B = 0·0     B = 0·2     B = 0·4     B = 0·6

Y = 0·2     Y = 0·2     Y = 0·2     Y = 0·2     Y = 0·2
R = 0·8     R = 0·6     R = 0·4     R = 0·2     R = 0·0
B = 0·0     B = 0·2     B = 0·4     B = 0·6     B = 0·8

Y = 0·0     Y = 0·0     Y = 0·0     Y = 0·0     Y = 0·0     Y = 0·0
R = 1·0     R = 0·8     R = 0·6     R = 0·4     R = 0·2     R = 0·0
B = 0·0     B = 0·2     B = 0·4     B = 0·6     B = 0·8     B = 1·0

Red                                                       Blue        Fig. 16

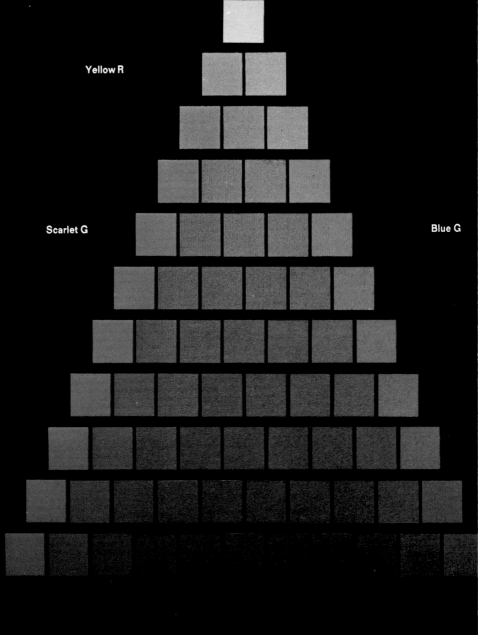

**Trichromatic Pattern Card**

# The mordant process for the natural dyestuffs 5

The mordant process must be first carried out before applying the natural products and dyes outlined in Chapter 6. Mordants are metal salts, such as stannic chloride, copper sulphate and the others that will be described in this chapter. While it is not necessary to mordant yarn when applying the commercial dyes discussed in the last chapter, the mordants are necessary to enter into chemical combination with the natural dyes to form insoluble 'colour lakes' which not only provide the combination necessary for the colour development, but also gives the dye complex an insolubility inside the fibre, rendering the colours more wash- and light-resistant.

The basic chemicals used in this chapter are as follows:

*Wool Dyeing*
| | |
|---|---|
| Aluminium mordant | Aluminium sulphate |
| | Cream of tartar |
| Tin mordant | Stannous chloride |
| | Oxalic acid |
| Chrome mordant | Sodium or Potassium dichromate |
| | Cream of tartar |
| Iron mordant | Ferrous sulphate |
| | Cream of tartar or |
| | Oxalic acid |
| Copper | Copper sulphate |

*Silk Dyeing*
| | |
|---|---|
| Aluminium mordant | Aluminium potassium sulphate |
| | Sodium thiosulphate |
| Tin mordant | Stannic chloride |
| Chrome mordant | Basic chromium chloride |
| | Sodium silicate |
| Iron mordant | Not applicable |
| Copper mordant | Copper sulphate |

*Cotton and Linen*
| | |
|---|---|
| Aluminium mordant | Tannic acid |
| | Aluminium sulphate |
| | Sodium carbonate (Soda ash or washing soda) |

Fig. 17

| | |
|---|---|
| Tin mordant | Tannic acid |
| | Stannic chloride |
| Chrome mordant | Chrome alum |
| | Sodium hydroxide (caustic soda) |
| Iron mordant | Tannic acid |
| | Ferric sulphate |
| Copper mordant | Tannic acid |
| | Copper sulphate |

For the home dyer, these salts may be purchased from chemical supply houses. Details of these are listed at the end of the book.

## COMMON SENSE WITH CHEMICALS

The metal salts are poisonous if ingested orally, but with care and attention, they are no more dangerous or harmful than other preparations found in the home. Keep the bottles well labelled and away from small children. When processing them at boiling temperatures keep the room well ventilated and do not allow the chemicals near food. Keep separate containers for the wet processing operations distinct from containers used in food preparation.

Glass stirring rods are best for controlling the yarn in dyeing and mordanting operations. Wooden spoons can be used as long as they are well washed when used for different operations. When the mordanting process is finished, follow the instructions relating to washing and drying operations.

## CALCULATION OF AMOUNTS

The recipes are given for the preparation of 50 g of textile material and usually a 30:1 liquor ratio. The figures in brackets after the chemical name are the amounts used expressed in percentage form. (For details of this calculation, refer to Chapter 1.) All quantities calculated from powder and *not* 10% solution.

### Mordant application on to wool

*Aluminium*
Recipe for 50 g    L.R. = 30:1

Add to the mordant bath the following

| | | |
|---|---|---|
| Aluminium sulphate | (6%) | 30 g |
| Cream of Tartar | (6%) | 30 g |

Stir to dissolve chemicals.

Add the wool and bring to the boil over 60 minutes and continue to boil for 30 minutes. Leave the wool in the bath and allow to cool, rinse in warm running water for 5 minutes and then prepare a fresh bath of hot water. Add the mordanted wool and boil for 20 minutes. Finally, squeeze dry and wrap in a towel until ready to apply the dye extract.

## Tin

Recipe for 50 g    L.R. = 30:1

Add to the mordant bath the following

|  | | |
|---|---|---|
| Stannous chloride | (4%) | 2 g |
| Oxalic acid | (2%) | 1 g |

Stir to dissolve the chemicals.

Add the wool and bring to the boil over 60 minutes and continue to boil for a further 60 minutes. Leave the wool in the cooling bath then rinse under warm running water for 15 minutes. Wash the wool in warm water and household detergent for about 15 minutes at 30°C. Rinse again under a warm tap then squeeze gently and wrap in a towel ready for the dyeing operation.

Note that the tinned wool must be washed in a warm wash solution after the mordant operation or it will become brittle and harsh.

## Chrome

Recipe for 50 g    L.R. = 30:1

Add to the mordant bath the following

|  | | |
|---|---|---|
| Sodium or potassium dichromate | (3%) | 1·5 g |
| Cream of Tartar | (3%) | 1·5 g |

Stir to dissolve the chemicals.

Add the wool and bring to the boil over 60 minutes and continue to boil for a further 60 minutes. Leave the wool in the cooling bath and then rinse in warm running water for 15 minutes. Squeeze gently and wrap in a towel.

Note: Wool treated with this mordant will turn green on exposure to light. For the best conditions and to minimise this shade change, mordant the yarn in the evening and out of direct sunlight. After rinsing, store in a light-proof cupboard.

Keep the room well ventilated and do not inhale the fumes from the boiling mordant bath. After rinsing the treated wool is harmless and non-toxic.

## Iron

Recipe for 50 g    L.R. = 30:1

Add to the mordant bath the following

|  | | |
|---|---|---|
| Ferrous sulphate | (5%) | 2·5 g |
| Cream of Tartar OR | (6%) | 3·0 g |
| Oxalic acid | | |

Stir to dissolve the chemicals.

Add the wool and bring to the boil over 30 minutes and continue to boil for a further 45 minutes. Remove the wool from the bath while hot and wrap immediately in a towel and keep wrapped for *three days* to complete the reaction. Then rinse the wool in warm running water, squeeze dry and wrap in a towel until ready for dyeing.

*Copper*

Recipe for 50 g of wool    L.R. = 30:1

Add to the mordant bath the following

                Copper sulphate                     (3%)      1·5 g

Add the wool to a warm bath and bring to the boil over 30 minutes and continue to boil for a further 30 minutes. Leave the wool in the cooling bath and then rinse under a warm running tap for 15 minutes. Squeeze gently and wrap in a towel until ready to dye.

## Mordant application on to silk

Apart from the chrome mordant, silk is treated differently from wool. Remember that the silk threads damage easily and must be treated in a careful manner.

*Aluminium*

Recipe for 50 g    L.R. = 30:1

Add to cold water (1·5 litres), the following

                Aluminium potassium sulphate    150 g
                Sodium thiosulphate               60 g

Stir to dissolve the chemicals.
Add silk and work for 60 minutes cold.
Increase the temperature to 40°C and hold for 60 minutes.
Increase the temperature to 60°C and hold for 40 minutes.
Increase the temperature to 80°C and hold for 40 minutes.
Cool the bath slowly then squeeze excess mordant from the hanks, rinse in warm water for 10 minutes, and then dry on a towel.

*Tin*

Recipe for 50 g    L.R. = 30:1

Add to cold water (1·5 litres)

                Stannic chloride                       375 g

Stir to dissolve the chemicals.
Add silk and soak overnight in the cold bath. Squeeze the excess mordant and rinse in cold running water for 30 minutes, then a warm rinse for 45 minutes. dry in a towel.

*Copper*

Recipe for 50 g    L.R. = 30:1

Add to cold water (1·5 litres)

                Copper sulphate                   1·5 g

Stir to dissolve chemicals.

Add silk and raise to 85°C over 30 minutes and hold for 30 minutes. Leave silk in a cooling bath, then squeeze when cold, rinse under warm water for 15 minutes. Wrap in a towel to dry.

Chrome
Recipe for 50 g    L.R. = 30:1
This is a two-bath process. For ungummed silk only.
Bath 1
Add to cold water (1·5 litres)

     Basic chromium chloride   30 g

Stir to dissolve chemicals.
Add the silk and soak for 6 hours, remove and squeeze gently to remove excess mordant solution.
Bath 2
Add to cold water (1·5 litres)

     Sodium silicate   30 g

Stir to dissolve chemicals.
Add the pre-treated silk and soak for 15 minutes. Remove and squeeze gently to remove excess silicate solution, leave on a towel for 2 hours, then rinse in warm running water for 30 minutes. Dry on a towel.

Iron mordanting will be covered in Chapter 6 when we talk about Logwood application.

**Mordant application on to cotton or flax**
Except for the chrome mordant, the preparation of cotton hanks depends on treating them with tannic acid. This chemical is necessary to fix the metallic salts on to the cotton fibres and to improve the wash fastness of the resulting dye/metal complex.
*Tannic Acid Impregnation* (common to the tin, iron and copper mordants).
Recipe for 50 g of cotton or flax  L.R. = 30:1
Add to cold water (1·5 litres)

     Tannic Acid (5%)   75 g

Add the cotton hank and raise to the boil over 20 minutes and hold at the boil for 30 minutes. Leave the cotton in the cooling bath overnight. Next morning squeeze out the excess tannin and rinse the cotton for 15 minutes in warm running water.
*Tin*
Recipe for 50 g of tannic acid impregnated cotton or flax.  L.R. = 30:1
Add to cold water (1·5 litres)

     Stannic chloride   150 g

Add the cotton hank and work in the solution for 6 hours, then gently squeeze and rinse in warm running water for 30 minutes. Wrap in a towel until ready for dyeing.

*Iron*

Recipe for 50 g of tannic acid impregnated cotton or flax.     L.R. = 30:1
Add to cold water (1·5 litres)

        Ferric sulphate        75 g

Add the cotton and work in the solution for 6 hours. Treat as for tin mordant.

*Copper*

Recipe for 50 g of tannic acid impregnated cotton or flax.     L.R. = 30:1
Add to cold water (1·5 litres)

        Copper sulphate        75 g

Add the cotton and work in the solution for 6 hours. Treat as for the tin mordant.

*Chrome*

There are various methods for treating cotton with chromium, but unfortunately they are all complicated and messy. The most successful method was devised by Horace Koechlin and is termed the alkaline chrome mordant method.

Recipe for 50 g of cotton     L.R. = 30:1
Add to cold water (1·5 litres)

        Chrome alum        300 g
        Caustic soda (32% solution)        750 ml

Add the cotton and work in the solution for 40 minutes, and without squeezing, wrap in an old towel for 24 hours. Then boil in plain water for one hour, rinse well in warm running water and store wet in a towel until ready for dyeing.

*Care.* The caustic soda solution is hazardous and should not be allowed to come in contact with the skin in its strong solution form. Wear gloves and keep small children away from the operation. Do not use aluminium pots. For those students who have access to hydrometers, a 32% solution of caustic has a specific gravity of 1·35 or 37·5° Baumé.

Take extreme care when dissolving caustic soda flakes in water. The reaction of the flakes with water produces heat capable of boiling the solution. Add the flakes slowly with careful stirring. If at all possible, mix the solution in a cold water bath to minimise the heat build up.

*Aluminium*

The following process is the main mordant sequence for the old Turkey Red process used extensively until the early 1900's for the production of Madder and Alizarin red shades. Because the production of the Turkey Red oil is difficult, the process has been modified and will rely on the tannic acid impregnation. For those people who wish to use the original Turkey Red oil process, it is suggested that they contact the dye supply companies direct and enquire whether the product is still available. Some writers have suggested an alum treatment before the tannic acid and this is the method practised by most craft dyers. The aluminium salts fix far better when the tannic acid is already present, and so in this present description, we will fix the tannic acid as in the previous iron, copper and tin mordants.

Recipe for 50 g of cotton.    L.R. = 30:1

Apply the tannic acid as described in the tin mordant section for cotton but do not rinse. Wring out the cotton and dry in an oven set at medium or 165°C (330°F). Take care not to scorch the cotton during the drying process. The material can be dried by ironing, again taking care that the cotton is not damaged, as you will find that the yarn is extremely brittle at this stage and must be handled with care.

PREPARATION OF BASIC ALUMINIUM SULPHATE BATH

*Solution 1*

Dissolve in cold water (1·2 litres)

| Aluminium sulphate | 240 g |

*Solution 2*

Add slowly to cold water (300 ml)

| Sodium carbonate | 40 g |

Make sure all the sodium carbonate is dissolved. It is a good idea to sprinkle the carbonate on top of the 300 ml of cold water and stir in slowly, as this chemical is hard to dissolve.

Next, add the carbonate solution slowly in 50 ml portions to the aluminium sulphate solution, stirring and dissolving the white precipitate that forms before adding the next portion of carbonate solution. When all the carbonate solution has been dissolved, stir for a further 5 minutes.

Take the tannic acid impregnated, dried cotton and work in the basic aluminium sulphate solution for 2 hours, stirring gently. Then remove the hanks, rinse and dry. Wash in a detergent, scour at 40°C for 5 minutes, rinse again and dry.

The dyeing process involves Alizarin paste, and is discussed in the next chapter.

# 6  Dyeing wool and cotton with natural products

**Vegetable extracts**
In the heading, the term natural products is used in preference to natural dyes, as there are very few natural dyes, and most of them are unavailable in Australia and New Zealand. The standard reference for dyes, The Colour Index, lists natural dyestuffs, and these are mainly grown in Asia, the West Indies and parts of Europe.

Therefore we can safely say that Australia is not richly endowed with native plants that could be classified as natural dyes, although it is possible to obtain a colour with almost any type of weed, flower, leaf, root or nut when dyed on a mordanted yarn sample. The possible range of yellows and browns, fawns and dull greens are unlimited, but for special effects this narrow choice soon becomes boring unless one can find the bright reds, oranges, greens and blues. To obtain these colours one has to travel slightly further than Australia, and look for animal and shellfish life, as well as the native plants of Asia and China. Examples of these colours are

          Reds        Cochineal
          Blues       Indigo
          Orange    Persian Berries
          Purple     Shellfish

Nevertheless the Australian bush and the city vegetable and flower gardens are latent storehouses for colour.

Just because no recipe exists for a particular flower extract does not mean that it will not work, and there is much to discover through experimentation. Once you have mastered the basic concepts there is nothing that cannot be tried. Be prepared to uncover the subtle tonings that exist in the world of nature, the colours that may not be evident in the visual appearance of the flower or weed. Because a flower is blue does not mean that it will dye a blue colour on wool or cotton. Chances are that it will be a fawn or yellow. There has not been enough experimentation in Australia or New Zealand to confirm categorically all possible effects from all possible flowers. Remember, dyeing was carried out using natural products from the dawn of history up until 1856, when Perkin discovered the first synthetic dyestuff, and the rough laws of chemical constitution of plants to colour produced became fairly easy to predict. In short, the initial comments still hold good: for the reds, oranges, blues and bright greens, one has to look further than Australia.

Only a few recipes are given below that you can follow, and there are good reasons for this. The first reason is that dependent on the climate and time of year, soil type and a host of other variables, the colour produced from, say a dockweed, will vary from one locality and time to another. The colour produced in May from docks gathered from a garden in Victoria may be different when gathered from a garden in New South Wales or Dunedin. The second reason is the wool, or to a less extent, cotton you use, may be different from the previous material.

PREPARATION OF THE COLOUR BATH

The natural products can be extracted from the plant of your choice by steeping in water and then simmering. Most recipes specify a time needed at the boil, but this will vary accordingly with the articles you have chosen. In some cases, too harsh a boil or too long a simmer will destroy the active ingredients, in other cases too little boiling will not extract enough. As with any other method, the colour of the bath will give you a good guide but be prepared to experiment. (figure 18)

Wherever possible use fresh flowers, leaves and twigs and do not forget to document all details in a notebook. As stated above, the active chemical composition will vary dependent on the time of the year and the climate, and a rather good exercise in dyeing is to sample a particular plant throughout the year and discover the changes that occur every month or so.

Fig. 18

*General Methods for Seed Pods, Bark, Roots, Husks and Nuts*

The generally accepted ratio of material to use is 1 kg of dry material to 500 g of wool that you will eventually dye, Gather the nuts or bark and soak overnight in about 15 litres of cold water. (*figure 19*)

Basic recipe for 500 g of wool

| Vegetable matter | 1000 g (1 kg) |
|---|---|
| Cold water | 15 litres |

Soak the vegetable matter overnight, then extract the natural colouring matter by boiling for 1 to 2 hours. Strain out the residue and cool. Place the **premordanted** yarn in the liquid and raise to the boil over about 15 minutes. Simmer for about 30 minutes and check every 10 minutes at the boil for colour development. It is a good idea to snip little bits of wool from the pattern and label them every 10 minutes. When the colour has reached the desired shade, cool the dyebath evenly and rinse the dyed wool under warm running water. Dry out of direct sunlight.

*General method for weeds, berries, leaves and flowers*

The ratio is increased to 1·5 to 2 kg of vegetable matter to 500 g of wool. Gather the material fresh, and soak overnight in about 15 litres of cold water.

Recipe for 500 g of wool

| Vegetable matter | 1500 – 2000 g (1·5 – 2 kg) |
|---|---|
| Cold water | 15 litres |

Soak the vegetable matter overnight, then extract the natural colouring matter by boiling for 30 minutes. Strain the residue and cool. Place the **premordanted** yarn in the liquid and raise to the boil over about 15 minutes. Simmer for 30 minutes, and check every 10 minutes at the boil for colour development. When the correct shade has been established, cool the bath and rinse the dyed wool in warm running water. Dry out of direct sunlight.

GENERAL HINTS ON DYEING

After the rinsing operation, the material may feel harsh and rough due to the metal mordants in combination with the tannins and other complexes formed in the dyeing operation. Softening with a commercial softener before drying will overcome this harsh handle.

SPECIAL COLOURING MATTER DERIVED FROM ANIMALS

*Cochineal*, the female of the *Coccus cacti*, is an insect which lives and propagates on certain types of cacti, especially the Nopal or *Cactus opuntia*. The plants and cacti are native to Mexico and Guatemala, but have been successfully introduced into the Canary Islands, Algeria, Indonesia and certain parts of Australia. The male insect is much smaller than the female, is furnished with wings, and does not yield the dye cochineal.

Fig. 19

The female remains attached to one spot on the plant, and at the age of three months is swept away from the leaves into straw baskets and killed by being thrown into boiling water and afterwards dried in the sun. The alternative is to place the insects in a bag and place them in an oven and it is this method that produces the so-called Silver Cochineal. About 70,000 insects are needed to produce 500 g of colouring matter, and an acre will produce about 500 to 600 kilograms of extract.

The active colouring matter is carminic acid and in the dyeing of wool, two distinct red shades are obtained, a crimson and a scarlet. There was a large dyeing industry built on this colouring dye, and at one time the uniforms for the English Army were dyed in this way. The dye is fast on wool and silk.

Cochineal gives the following series of shades with the various mordants:

| | |
|---|---|
| Chromium | Purple |
| Aluminium | Crimson |
| Iron | Purple |
| Copper | Claret |
| Tin | Scarlet |

Tin and aluminium salts are the fastest mordants used in cochineal dyeing of wool and silk.

DYEING WOOL WITH COCHINEAL. L.R. 30:1

The dyeing may take place on pre-mordanted material, or mordanting and dyeing in the same bath, and it is the latter that is usually adopted. All quantities as for Standard Formula for undiluted dyes and chemicals. See chapter 1.

*Separate Mordanting.*
Either use the pre-mordanted yarn prepared in the previous chapter, or follow this recipe
For 500 g of wool:

| | | |
|---|---|---|
| Stannous chloride | 5% | (25 g) |
| Cream of tartar | 5% | (25 g) |

Immerse the wool at 60°C and gradually raise to the boil over 30 minutes. Simmer for 45 minutes and then cool the bath. The material can then be either washed in a detergent bath or taken direct to the cochineal dyebath composed of

| | | |
|---|---|---|
| Cochineal powder | 10% | (50 g) |

Continue boiling for 90 minutes.
*Crimson Dyeing*
for 500 g of wool

| | | |
|---|---|---|
| Alum | 6% | (30 g) |
| Cream of Tartar | 6% | (30 g) |

Immerse the wool in the mordant bath at 60°C and gradually raise to the boil over 30 minutes. Simmer for 45 minutes and cool the bath. Transfer to the cochineal bath composed of

| | | |
|---|---|---|
| Cochineal powder | 20% | (100 g) |

Continue boiling for 60 minutes.
*One-bath dyeing*
for 500 g of wool
Add to the dyebath in the following order:

| | | |
|---|---|---|
| Oxalic acid | 5% | (25 g) |
| Stannous chloride | 4–5% | (20–25 g) |
| Cochineal powder | 10–20% | (50–100 g) |

Without the yarn, raise the dyebath temperature from 60°C to the boil in 30 minutes and boil for about 10 minutes. Cool the bath down with cold water and enter the wool at 70°C. Raise to the boil in 45 minutes and boil for 45 minutes. A deficiency in tin gives dull bluish shades, while an excess gives a pure scarlet. Cream of tartar increases the intensity, and if part of the oxalic acid is replaced on a 50/50 basis with cream of tartar, the colour develops into an orange, the more tartar being used, the more yellow the dyeing will become.

The one-bath system gives brighter and yellower shades than those obtained by using pre-mordanted yarn. To produce the variations of clarets and crimsons, small amounts of iron or chrome salts can be added to the one-bath method. Purples and crimsons can be produced on pre-mordanted yarn by adding a little alkali in the form of ammonia to the bath where alum or tin pre-mordanted yarn has been used.

*Preparation of Ammoniacal Cochineal*

Cochineal powder and ammonia are allowed to stand for 4 days. The proportions are

        1 part cochineal powder
        3 parts concentrated ammonia solution

The mixture is heated in a very well ventilated room and the excess ammonia is driven off with heat. Add to the resulting mass half of its own weight of aluminium oxide and heat further to drive off all the free ammonia. When the mass has cooled down, mould the powder into cakes and store in a dry place. This ammoniacal cochineal is used for dyeing wool crimson and purple, and in conjunction with ordinary cochineal, for rose reds and pinks.

*Recipe for a rose pink on wool (figure 20)*
For 500 g of wool

| | | |
|---|---|---|
| Ammoniacal cochineal powder | 2% | (10 g) |
| Cochineal powder | 2% | (10 g) |
| Tin metal | 1% | (5 g) |
| Nitric acid (**handle with care:** corrosive acid) | 8% | (40 ml) |

Dissolve the tin in the nitric acid, and then prepare the dyebath as follows: Add the ammoniacal and the ordinary cochineal to water at 50°C. Add the dissolved tin solution and make up the bath to the correct liquor ratio. Add the wool and raise to the boil over 30 minutes. Simmer for 60 minutes, cool and rinse the wool in warm running water for 60 minutes. Wash in a detergent solution and then soften in a commercial softener.

GENERAL NOTES ON COCHINEAL DYEING
If the cochineal powder cannot be obtained, it is possible to use the cochineal solution sold in the supermarkets. The strength varies from one manufacturer

Fig. 20

to another, but as a general rule estimate that the given solution is about 10% strong. Therefore use 10 times as much. Most wool that has been dyed in the presence of tin salts will be harsh and inclined to be brittle unless the un-reacted tin salts are removed from the finished wool. This can usually be accomplished by rinsing the dyed material in warm running water for about one hour. Softening with commercial softeners will also help to mask the rough handle. If the harshness is still present after this treatment, a rinse in a weak solution of Calgon should help. Use care in this treatment, as too much Calgon will destroy the colour. Normally, a solution containing 0.5 gm per litre will be sufficient at a temperature not exceeding 30°C for 20 minutes.

DYEING WOOL WITH KERMES

Kermes is similar in shade to cochineal and also comes from a dried insect known as *Coccus illicus* which live in a species of oak, the *Quercus coccifera*. Kermes is still used in Asia and some parts of Europe in the dyeing of wool and leather. The application is similar to that of cochineal.

DYEING WOOL WITH LAC

Similar to the previous animal dyes, lac is a product of the insect *Coccus lacca*, and is a secondary product in the manufacture of shellac. Lac is faster

than cochineal dyeings but not as brilliant, and is used in the production of scarlets, oranges and crimsons on wool and silk.
Recipe for a lac scarlet on wool — for 500 g of wool.

    Lac powder    8%
    Hydrochloric acid (**handle
    with care:** corrosive acid) 4%
    Cream of tartar   5%
    Stannous chloride  5%

Dissolve the lac powder in the hydrochloric acid and stannous chloride overnight. Bring the bath volume to the correct liquor ratio, add the lac/tin solution and the cream of tartar, and bring to the boil. Boil for 10 minutes, cool to 70°C with cold water and then add the wool. Raise to the boil over 30 minutes and then simmer for 45 minutes. Cool bath, rinse and dry.

DYEING WOOL WITH LICHEN AND MOSS EXTRACTS
Lichens, for the non-biologists, are fungus and algae that live together in a kind of partnership called symbiosis. The algae contains chlorophyll and so can produce food that it shares with the fungi. The fungi in turn absorbs water that helps keep the algae moist. Lichens grow chiefly on dead tree stumps and rocks. They range in colour from yellows, oranges, fawns and browns. There are about 15,000 different types of lichens and they can be found in any climate. The chemical indicator, litmus is prepared from lichen growth.

Orchil and Cudbear are colouring compounds derived from various lichen families, and belong to a fairly small class called Cryptogamia. Commercially they were sold under the name of Orchella weed, Orchil weed, Valparaiso weed and Lima weed. Lichens do not contain strong colouring matter already formed, but certain colourless compounds, which, by the action of ammonia and oxygen, are converted to colouring matter.

EXTRACTION OF COLOURING MATTER BY FERMENTATION WITH AMMONIA.
The lichen is picked and shredded, placed in bottles and covered with a dilute solution of ammonia, 1 part of lichen to 3 parts of ammonia solution. Keep the temperature of the mass about 35-45°C for about 5 or 6 days. When the fermentation commences, watch for the colour development by testing with some white wool. If a dirty brown colour starts to develop, the fermentation has progressed too far and must be stopped at once. (figure 21)

The orchil paste produced will dye wool to a magenta colour in a neutral bath containing no assistants. Addition of small amounts of oxalic or sulphuric acid will turn the shade brighter and redder.

CUDBEAR
Cudbear is a similar dye to orchil and is prepared from the lichen Lecanora tartarea. It can also be produced commercially as a brownish purple powder.

Fig. 21  The powder can be produced in the same manner as above.*

Experimenting with the different varieties can be a fascinating experience, once the lichens are found. The growth can be scraped from the log or rock and a simple test performed to see if the Lichen contains the active ingredients. Milner describes testing the thallus of the growth with a needle dipped in a weak sodium hypochlorite solution. If the active colouring acids are present, the drop turns red. The selected lichen can then be scraped off the support and placed in a labelled bottle describing the location and time of year.

### Dyeing Wool with Special Dye Woods
FUSTIC

This dyestuff is sometimes known as Old Fustic, Cuba wood and Yellow wood. It comes from the wood on the trees belonging to the Urticaceoe known

---

* Anne Milner in her book NATURAL WOOL DYES AND RECIPES (John McIndoe, NZ 1971) presents many recipes based on crustaceous lichens.
Other books dealing with this interesting subject are: *New Zealand Lichens* William Martin and John Child A.H. Reed, 1972 and *Lichens for Vegetable Dyeing* Eileen Bolton-Robin & Russ.

botanically as *Morus tinctoria* or *Maclura tinctoria*. It is a native of Brazil, Mexico and some of the West Indian Islands.

It is used as chips, rasped wood or as a paste. Before the introduction of the synthetic dyes, it was the most important yellow colouring pigment for wool, and was used extensively in conjunction with logwood for black dyeing and with other natural colouring materials such as indigo and madder to produce combination shades of browns, olives and drabs. In most cases the mordant used is sodium dichromate, and these produced, with the other natural dyes, the dark colours. With tin or aluminium mordants, the colour was yellow, but turned browner as exposure to light changed the chemical complex. Fustic gives the following series of shades with the various mordants

| | |
|---|---|
| Chromium | Olive yellow to brownish yellow |
| Aluminium | Yellow |
| Iron | Dark olive |
| Copper | Olive |
| Tin | Bright yellow to orange yellow |

The heartwood of the Cockspur Vine or Thorn, *Cudrania javanensis*, growing in New South Wales or Queensland, is the nearest available to fustic. The fustic extract is commercially available in Britain, although some importers are reported to have carried stocks in Australia.

*Recipe for single bath dyeing of Fustic Yellow*
For 500 g of wool.

| | | |
|---|---|---|
| Stannous chloride | 4% | (20 g) |
| Oxalic acid | 4% | (20 g) |
| Fustic extract | 20% | (100 g) |

Start the dyeing cold with the ingredients and raise to the boil in 30 minutes. Simmer gently for 30 minutes, cool and rinse in warm running water for 30 minutes.

QUERCITRON BARK AND FLAVINE

The colouring matter contained in the inner bark of the oak species *Quercus nigra* or *Quercus tinctoria*, a native of America. The bark is removed from the tree, dried and ground between millstones and finally appears as a mixture of woody fibre and a fine yellow buff powder. Flavine is a preparation obtained from the bark and is reportedly twenty times stronger than Quercitron.

*Preparation of Flavin*

| | |
|---|---|
| Quercitron bark, shredded and ground | 100 parts |
| Water | 300 parts |
| Sulphuric acid (concentrated) | 15 parts |

Boil the solution for 4 hours, cool and filter the paste. Wash free from acid and then dry. The mordants used with the Fustic system will produce similar colours.

Fig. 22

*Persian Berries, Cochineal, Fustic with the five mordants*

WELD
The colouring matter contained in the species *Reseda luteola* which was once cultivated in England, France and Germany. It is a herbaceous plant that grows to 3 feet and the upper part of the plant, especially the leaves and seeds, contain the active colouring matter. Weld is applied in a similar manner to Fustic and Flavin and gives the following series of shades with the various mordants:

| | |
|---|---|
| Chromium | Olive yellow |
| Aluminium | Greenish yellow |
| Tin | Bright yellow |
| Iron | Olive |
| Copper | Yellow olive |

PERSIAN BERRIES
These are the fruits of the buckthorn and species *Rhamnus*, cultivated in southern Europe and Asia. The berries of *Rhamnus amygdalinus* yield the best quality.

Persian berries, like quercitron bark, are used principally in the form of extracts, and were used in considerable quantities for the production of yellows, oranges and greens on cotton printing. In wool dyeing, the yellows turn brown on exposure to light and do not appear to have any advantage over Flavin or Weld. Used in conjunction with cochineal, they will produce oranges and scarlets.

Persian berries give the following series of shades with the various mordants:

|           |                |
|-----------|----------------|
| Chromium  | Brown          |
| Aluminium | Bright yellow  |
| Tin       | Orange         |
| Iron      | Dark olive     |
| Copper    | Yellowish olive|

Other Types of Natural Dye Products.

| Botanic Name | Origin | Description |
|---|---|---|
| *Rhus cotinus* | West Indies | Similar to Flavin |
| *Curcuma tinctoria* (Saffron) | China and India | No mordants necessary to dye cotton and wool. Non-fast to light and washing. |
| Safflower *(Carthamus tinctoria)* | Egypt and India | Cotton dye, non-fast |
| Soluble Red woods: | Brazil | Brazil wood |
|  | Mexico | Peach wood |
| Genus *Caesalpinia* | Thailand ⎫ | Sapan wood |
| Order Leguminosoe | Japan ⎬ |  |
|  | Peru ⎭ | Lima wood |
| Insoluble Red woods: |  |  |
| *Pterocarpus santalinus* | India, Ceylon | Saunders wood |
| *Baphia nitida* | Sierra Leone | Barwood |
| unknown | African west coast | Camwood |

MADDER (ALIZARIN)

The root of the plant *Rubia tincorum*, a native of Asia minor and used as a dye on cotton since Roman times. With the various mordants, the colours are:

|           |                      |
|-----------|----------------------|
| Chromium  | Crimson              |
| Aluminium | Pink to scarlet      |
| Iron      | Maroon to red-brown  |
| Copper    | Yellow-brown         |
| Tin       | Dull red-orange      |

Madder was used to dye French military uniforms, and in China to dye carpets, but the biggest use appears to have been on cotton by a similar process to the Turkey Red process. Madder is now called Alizarin.

ALIZARIN ON WOOL AND COTTON

Synthetic alizarin can be obtained from most dye suppliers and is sold as a paste containing about 20% of the dyestuff. Like cochineal and lac, it is polygenetic, giving the following colours with the different mordants:

| | |
|---|---|
| Aluminium | Red |
| Tin | Pink |
| Iron | Violet |
| Chromium | Puce brown |
| Copper | Yellow-brown |

To dye alizarin on cotton you should use the tannic acid/aluminium mordant yarn. On wool it is generally dyed using aluminium or chromium mordants. The presence of calcium acetate or other soluble calcium salts is an essential condition to obtaining good colours with alizarin due to the formation of the double calcium and aluminium lake.

*Application to wool.* (Using pre-mordanted wool.)

Take the alum or potassium dichromate mordanted wool and immerse in a dyebath composed of:

| | |
|---|---|
| Alizarin paste (20%) | 10% |
| Calcium acetate | 4% |
| Stannous chloride | 2% |

The dyebath is raised slowly to the boil and kept at the boil for 30 minutes. Rinse and dry in the normal manner.

Brighter shades are obtained when the stannous chloride is added to the aluminium mordant bath. The colour produced has an orange shade dependent on the amount of stannous chloride present. If the wool is mordanted only with stannous chloride, a variation of the tin mordant, alizarin orange is produced. For this purpose from 5-8% of both stannous chloride and cream of tartar are used, and the mordanted goods are washed and dyed as above.

To mordant with potassium bichrome or dichromate use 3%, with or without 1% of sulphuric acid. Iron is used also as a mordant for alizarin on wool. From 4-8% of ferrous sulphate and 4-8% cream of tartar are used in the mordant bath.

*Application to cotton.*

Refer to the previous chapter under the section relating to aluminium mordant on cotton.

The pre-mordanted cotton is dyed from the following dyebath:

| | |
|---|---|
| Alizarin paste (20%) | 15% |
| Calcium acetate | 4% |

Immerse the cotton at 25°C and work in the bath at this temperature for 30 minutes. Slowly heat the bath to 60°C over 30 minutes and dye at this temperature for 30 minutes. Rinse and dry as normal. The cotton should be steamed for 2 hours. After steaming, wash the cotton in a 0.5% soap solution in water. Wash and dry.

A longer and more complicated dyeing can also be carried out using the following system which approximates the traditional turkey red process. Scour the cotton by boiling it with sodium carbonate (washing soda). Work in a bath containing 10 parts of turkey red oil and 90 parts of water. When saturated, squeeze and dry at a temperature of 40°C to 50°C. **The operation of turkey red oil and drying is carried out twice.**
Work the cotton in a cold bath containing 4% aluminium acetate then wring out and dry at between 40°C and 50°C. **Repeat this process.**
Work in a bath at 35°C consisting of

6 g per litre ground chalk for 30 minutes.

Wash well in water and then dye using

| | |
|---|---|
| Alizarin paste (20%) | 15% |
| Calcium acetate | 4% |

Start at 20°C and work for 20 minutes. Heat slowly at 60°C over 20 minutes. Continue dyeing at 60°C for 30 minutes. Steam and wash as shown in previous section.

## LOGWOOD ON WOOL

There are many recipes for the application of logwood extract on wool, perhaps the easiest method is the dyeing on pre-mordanted material. The following example indicates a chrome recipe but all mordants described in the previous pages will produce a coloration with this dye.
Recipe for 50 g of wool
Mordant the wool with

| | |
|---|---|
| Potassium dichromate | (3%) |
| Formic acid | (3%) |

Immerse the wool cold, raise to the boil in 15 minutes and boil for 45 minutes, rinse and then dye in a bath containing either 80% logwood chips or 10% Hematine paste or crystals.

## LOGWOOD ON COTTON

Refer to Mordant Application for Cotton.
Stannic chloride mordant, then 30% logwood solution — purple dye formed.
Copper sulphate mordant, then 30% logwood solution — Blue.

Many Australian dye companies will be able to supply the logwood chips or Hematine paste or crystals. It is recommended that readers who wish to experiment with logwood on natural fibres should contact the dye companies for their recipes. Most of the old recipes are lengthy and involve detailed preparation, multiple impregnation and coloration processes, and are considered outside the scope of this book.

**Mineral Colours**

These are not dyestuffs as such, but inorganic compounds, insoluble in water, and precipitated in the fibre by a suitable double-decomposition reaction. They have no affinity for textiles but are mechanically held.

Chrome yellow — cotton
Saturate the cotton in a cold bath containing

|  |  |
|---|---|
| 5% Lead acetate | 5 g/100 ml |

Squeeze and pass into a hot bath containing

|  |  |
|---|---|
| 5% Potassium dichromate | 5 g/100 ml |

Wash well in hot soapy water.
To convert in to an orange, pass the dyed cotton through a lime solution in water.
The chrome yellow and orange is very fast to light and washing.

Iron buff — cotton
Saturate the cotton in a cold bath containing

|  |  |
|---|---|
| 5% Ferrous sulphate | 5 g/100 ml |

Squeeze and pass into a hot bath containing

|  |  |
|---|---|
| 5% Sodium carbonate | 5 g/100 ml |

The cotton is then washed and exposed to air and sunlight for 6 hours.

Chrome green for wool and cotton
Saturate the wool or cotton in a cold bath containing

|  |  |
|---|---|
| 10% Chromium sulphate | OR |
| 10% Chrome alum | OR |
| 10% Sodium dichromate | 10 g/100 ml |

For cotton, squeeze and place in a boiling bath of

|  |  |
|---|---|
| Sodium carbonate | 5% solution |

For wool, squeeze and place in a warm to hot bath of

|  |  |
|---|---|
| 5% Sodium bisulphite | OR |
| Sodium metabisulphite | 5 g/100 ml |

Mineral khaki for cotton
Saturate the cotton in a cold bath containing

|  |  |
|---|---|
| 5% Ferrous sulphate | 5 g/100 ml |
| 5% Chrome alum | 5 g/100 ml |

Squeeze and pass into a boiling bath containing

        5% Sodium carbonate     5 g/100 ml

After washing, expose to the air to oxidise.

*Prussian blue for cotton*

Soak cotton in a warm (30°C) bath containing

        5% Ferric sulphate       5 g/100 ml
        2-5% Stannous chloride   2-5 g/100 ml

Squeeze and work in this solution for 20 minutes, then, after squeezing, transfer to a bath containing

        10% Potassium ferrocyanide  10 g/100 ml
        2% Sulphuric acid         2 g/100 ml

Work cotton for 30 minutes, then rinse and dry.

*Prussian blue for wool*

Soak wool in a warm (30°C) bath containing —

        10% Potassium ferricyanide  10 g/100 ml
        2% Sulphuric acid **(handle with care)**     2 g/100 ml

Work for 15 minutes at 30°C and then raise gradually to boiling point. Boil 10 minutes, rinse and dry.

The addition of 1-5% of stannous chloride (based on the bath volume and not on weight of wool) to the bath either at the beginning or at the end, gives a purple shade on the wool.

Alternatively the process can be carried out on tin mordanted wool.

# 7 Special Effects

Up to the present the processes described have been traditional production processes and there are many methods that have been used over the years to introduce variations. Most have been simple variations, while some are intricate and depend on specialised machinery to produce the effect. In the following section, three methods are described where the dyeing system has been modified to produce multicoloured effects.

Methods are also shown for cold dyeing where the dye solution is squeezed into the textile material and left for a day to react with the cotton or wool. This is an important part of commercial dyeing operations where colour variations are produced without resorting to traditional dyebath techniques, and the resulting range of variations are increased.

The last section in this chapter describes one method used to shrink-proof woollen garments or hanks. The process is simple and the ingredients are easily obtained.

**Dip Dyeing.**
The revival of ombre dyeing over the past few years has introduced many novel effects. It can be duplicated in most forms by the home dyer. The effects produced are graduations of similar, or different colours around the perimeter of the hank, so that when knitted or woven, a multicoloured effect is produced.

*Method*
Select a pale shade that you wish to reproduce on the final shade and dye this colour on the complete hank in the normal manner. After this has been done, prepare a new dyebath of a different colour, choosing one that will blend with the first colour produced. Suspend the hank into the new dyebath and dye only one quarter of the hank as shown in figure 23. Bring to the boil quickly and dye at the boil for 15 minutes. Repeat this process around the circumference of the hank using different dyebaths until the hank has the desired combinations.

Recommended dyes:

    *Wool*      Acid milling or acid levelling dyes.
    *Cotton*   Direct colours or reactives.

Fig. 23

## Tie Dyeing

This technique also produces a random effect on hanks or fabric, although fabrics dyed in this manner look more effective in final appearance. Lay the material, either hanks or fabric, on the table and bunch together as shown in figure 23. Tie string around the bunched sections and tie tightly. Next you must dye the material in the usual manner, calculating the liquor ratio, adding dyes and chemicals consistent with the type of material used, i.e. cotton or wool. At the conclusion of the process, untie the string and rinse well. Retie the string into different sections of the material and re-dye in a different dye solution.

Recommended dyes:

>*Wool*     Acid milling or acid levelling dyes
>*Cotton*   Reactive or Direct

## Batik Dyeing for Cotton

This method is practised in Indonesia where some beautiful art forms are produced.

A design is drawn on the fabric and those sections which are not to be dyed are coated with hot liquid paraffin wax. Prepare a cold reactive dyebath and dye the cotton in the usual way (see Chapter 5). At the conclusion of the dyeing process, boil the material with a type A detergent and the wax will liquify and float free from the material. The process can be repeated as many times as necessary to colour the patterns. By cracking the wax after it has been applied to the cotton, very fine lines can be produced.

Recommended dyes:

>*Cotton*   Reactive cold dyeing types

## Cold Pad Batch Techniques

The random effects produced by dipping or tie dyeing can also be produced using a technique of applying a cold, thickened solution of dye to the yarn, and storing cold for 24 hours. Greater variety can be produced by this method as it is possible to apply the colours not possible in the dip method. As you will see, the application is simple and direct, although special chemicals and auxiliaries are needed. The methods shown are for nylon and wool, and a recipe is also shown involving direct colours for cotton and flax.

RECIPE FOR COLD PAD BATCH ON WOOL AND NYLON

Preparation of one litre of solution. Wool reactive dyes.

|  |  | *Supplier* |
|---|---|---|
| Urea crystals | 300 g | |
| Wetting Agent | 2 g Nobosol OT | Croda Chemical Company |
| Acetic Acid (65%) | 3 ml | |
| Thickening agent | 10 g Manutex RS type | Albright & Wilson |
| Methylated spirits | 10 ml | |

## Method

1. Measure 500 ml of water at 40°C in a glass beaker and add urea.
2. In a separate beaker, wet out the Manutex RS with the methylated spirits and stir to remove lumps.
3. Add the Manutex/methylated spirits mixture to the dissolved Urea solution and stir rapidly until all the lumps have been dissolved.
4. Add the acetic acid and wetting agent and stir well, strain the mixture through a fine cloth and make up the volume to 1 litre.
5. Divide this mixture into 5 or 6 equal parts. Weigh up 5 or 6 different wool reactive dye powders and dissolve them separately in a minimum of hot water.
6. Add the dye solutions to the separate Urea/Manutex mix and stir till the dye is dissolved.
7. Wash the wool or nylon hank in warm water and squeeze dry.
8. Lay the hank over brown paper on a tray. (figure 24)
9. Pour the dye solution carefully over sections of the hank (figure 25) to give five or six bands of colour. When all dyes have been poured, place hank in an old towel and either wring or roll with a rolling pin along the length of the hank. (figure 26)
10. Remove the hank from the towel and place in a plastic bag and store for 24 hours at room temperature. (figure 27)
11. Wash in a warm detergent bath and dry.

Fig. 24

Fig. 25

Fig. 26

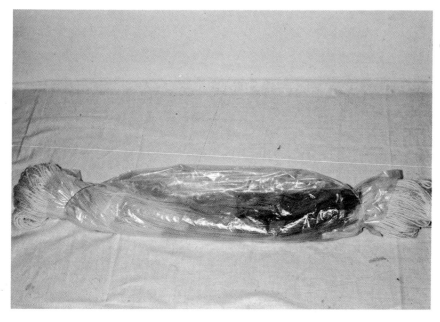

Fig. 27

RECIPE FOR COLD PAD BATCH ON COTTON
Preparation of *one litre* of solution using direct cotton dyes

| | |
|---|---|
| Urea crystals | 50 g |
| Wetting agent | 2 g Nobosol OT |
| Ammonium sulphate | 2 g |
| Thickener | 5 g Manutex RS |
| Methylated spirits | 3 ml |

Method 1
1  Repeat the process of mixing as for the wool cold pad batch recipe dissolving the urea in 500 ml of water at 40°C. Wet out the thickener with methylated spirits. Add the thickener to the urea, then the Ammonium sulphate and the wetting agent.
2  Direct dyes are dissolved in a minimum of hot water as before and added to the solution.
3  Colour the cotton in exactly the same way as the wool.

**Blow-out Dyeing**
A variation of the dip process can be achieved by saturating the cotton or the wool hank in its appropriate cold pad batch mixture, without the added dyestuffs, squeezing out the excess solution with a laundry wringer or a rolling pin. Next, move outside on a windless day and use six or eight different dye powders mixed together in their dry state on the end of a spoon or spatula. Hold the hank by the outstretched left hand, and hold the spoon in the right hand in front of the mouth. Gently blow the mixed powder onto the hanks. Wrap the hank in plastic and batch for 24 hours, then wash off. The effect is to have the hank coloured by a variegated blotches which can be attractive when knitted.

**Shrink Resist Treatment of Pure Wool**
One of the disadvantages facing the home spinner or dyer is the fact that the garment, if made from home spun wool, is not resistant to shrinkage after normal hand washing. Garments that have been produced by long hours of hand spinning and knitting do have a tendency to pill and rub, causing a sometimes undesirable surface effect, as well as a high shrinkage factor when washed.

Not all spinners and dyers will want to treat their garments to withstand the pilling and shrinking effect, but for those who do, the process is as simple as the straightforward dyeing process. The main restrictions are that the process may sometimes dull pale bright shades in the acid levelling or milling class. The method of testing is therefore to dip a small cutting from the yarn and subject this to the shrink treat bath. Yarn and fabric can be tested in the same way.

RECIPE FOR A SHRINK RESIST TREATMENT ON WOOL

| | |
|---|---|
| Swimming pool chlorine powder | 2 g |
| Wetting agent | 5 g |
| Common salt | 100 g |
| Sodium metabisulphite | 20 g |

These ingredients will make the two solutions necessary for the treatment. Recipe is for 50 g of wool. L.R. = 30:1

*Solution 1* (Stable for 2 hours)

Add to 1500 ml of cold water:

| | |
|---|---|
| Chlorine powder | 2 g |
| Common salt | 100 g |
| Wetting agent | 5 g |

Stir until dissolved.
*Solution 2* (Stable for 2 hours)

Add 20 g of sodium metabisulphite to 1500 ml of cold water and stir to dissolve the powder.

*Procedure*

1. Wash the garment or hank in warm water containing a little detergent. Squeeze or wring out the material.
2. Place the wool in solution 1 and stir gently for 30 seconds.
3. Remove from the solution and wring gently by hand or a laundry wringer.
4. Place the wool in solution 2 and stir gently for 30 seconds.
5. Rinse under warm running water for 15 minutes, soften with a commercial wool softener and dry.

Woollen goods treated in this way are resistant to a normal hand wash in normal wash detergents or soap. Because shrinkage resistance also depends on the knitting tension, which may or may not be optimum, especially where coarse counts are spun, the treatment may not be sufficient for gentle machine wash conditions. A further treatment exactly the same as shown above can therefore be given, and it is this second treatment that will give a good resistance to machine washing on a gentle cycle.

# 8 Tanning recipes for sheep skins

The tanning process involves the use of chemicals to process animal hides into leather. For the home or school dyer, the process is simple and direct. The recipes require similar chemicals to those used in the dyeing process, all of which can be purchased from the supply houses.

**Purchase of the Skins**
Stock agents, abattoirs and farmers will sell sheep skins to the public, the price dependent on the current price of wool being asked at auction. The skins should be fresh and from recently-slaughtered sheep and in no way subject to putrefaction. Do not be concerned with the blood and flesh that adheres to the hides as this will be removed in the first two operations. If you purchase skins you should attempt to tan them within two days, but if this is not convenient, salt them and store till ready.

**Salting Process**
Take the green skins and rub the flesh side with an amount of common salt equal to the weight of the skin. Rub into the skin with the fingers then fold the hide down the backbone, skin to skin, and store in a well ventilated, dark, fly-free environment until you are ready to commence the tanning operation. Under no circumstances store the hides in a plastic bag as they will putrify very quickly, even with the salt preserve.

**Description of the Process**
All hides are soaked to remove the blood and bacteria, fleshed to remove the fatty tissue, and then scoured to remove the wool grease and dirt from the fleece. The next stage in the operation is the pickling process where salt and acid swell the skins in preparation for the actual tanning operation.
There is a choice of tanning recipes that you can use after the pickling process and the four mentioned below are described in this chapter.

1    Alum tan, used for light tanning where water and heat will not be used to wash the skin
2    Vegetable extract and bark bush tanning recipes.

3   Formaldehyde tanning, again rather light and not washing resistant
4   Chrome tanning for wash resistant leathers

After the tanning, the goods have to be oiled, or fat liquored with a sulphonated oil, dried and then 'broken' to induce softness.

In all the following recipes extreme care should be exercised in the control of temperature. Do not allow the hides to come in contact with baths over 25°C and in all recipes check with the thermometer to ensure that the treatment solutions are about 25°C. Check also that the liquor ratio is about 45:1 and note that the chemicals used are given in concentration units of grams per litre.

SOAKING PROCESS
Prepare a cold solution of

1 g/litre home detergent
1 g/litre White King® bleach solution.

Soak the green skins for two hours to remove the blood and kill the resident bacteria. If you have salted your skins and stored them you should soak the skins for 24 hours. This process is also used as a preliminary treatment for other animal skins.

FLESHING PROCESS
This process removes the fatty tissue from the hides so that the tanning agents work direct on the leather and not on skin tissue. There is no easy way to perform this function which consists of scraping off the tissue with an old saw blade that has had the teeth filed down. Hacksaw blades are also used for this operation. The hide is thrown over a log, flesh side up, and the blades used to carefully scrape off the skin tissue. Do not cut the skins below the subcutaneous level otherwise the skin will be rendered useless.

SCOURING OPERATION
Now the hide must be scoured to remove the wool grease and dirt that the sheep has picked up around the farm paddocks. The operation must be carried out with care so that the wool does not matt, and make sure the temperature is below 25°C.
Make up a solution containing

2.5 g/litre of detergent.
(L.R. 45:1 on weight of skin)

Gently work the skin around the solution for 15 minutes, then drain and rinse well. Repeat the scouring and rinsing process three time always using a fresh solution of detergent. In the last rinse, run warm water (not over 25°C) over the skins for 30 minutes. The hides will still contain burrs and other vegetable matter, but ignore these until the skins have received all the tanning processes and then card out the fleece with a carding comb.

## PICKLING PROCESS
The hides must now be swollen prior to the actual tanning operation. Make up the following solution:

        60 g/litre Common salt
        4 g/litre Formic acid (85%)

Allow the skins to remain in the pickling solution for 24 hours, stirring every now and then to keep the solution fresh and active. Remove the skins and let them drain for about 10 minutes but **do not let them dry out.**

TANNING PROCESS – (Four different methods)
*Alum Tan method*
Prepare a solution containing the following

        60 g/litre Common salt
        60 g/litre Alum

Allow the skins to remain in this bath for 24 hours or longer, there is no damage that can be done, and many tanners advocate at least 48 hours. If the weather is cold then leave the hides in the solution longer. The next process after tanning is fat liquoring which will be described in the section following tanning.

*Vegetable Extract method*
There are a number of ways of obtaining tannic acid, either by purchasing the commercially available Mimosa extract or by collecting your own bark and leaching the tannic acid with hot water.
*Mimosa Extract.*
Prepare a solution containing the following:

        30 g/litre Common salt
        15 g/litre Mimosa extract

Add the skins and allow them to remain in this bath for 48 hours. After the first 18 hours start adding a solution of Mimosa extract until you have added the equivalent of 30 g/litre. Soak for a further 24 hours, drain and oil.

*Bush Tanning*
Collect the bark from a gum or a wattle tree, shred it into small sections and pour boiling water over it and allow to cool. Collect the dark coloured Tannic solution. Keep collecting the bark and leaching until you have enough volume to treat the hide. Add to the tannic acid solution 30 g/litre common salt and leave the skins in the solution for 2–3 days, depending on the required

colour. Drain and rinse well, oil and dry. (See *Oiling or Fat Liquoring Operation* below.)

*Formaldehyde Tanning Method*
Prepare a solution containing the following:

> 40 g/litre Common salt
> 10 g/litre Formaldehyde (40%) solution

Allow the skins to remain in this bath for two hours, remove and drain. Replace the skins and leave overnight. Then add a further 20 g/litre of common salt, leave for 3-4 hours and then add sodium bicarbonate solution until the pH is 7. Leave the skins for 4 hours, rinse, drain, oil and stretch.

*Chrome Tanning Method*
Prepare a solution containing the following:

> 300 g/litre Chrome alum

Heat the solution to 35°C to dissolve the crystals and then leave to cool. Next, prepare this solution:

> 150 g/litre washing soda.

This solution will have to be heated almost to the boil to dissolve the soda. Next add the cold soda solution slowly to the chrome alum solution stirring during the addition. Let the resulting basic chrome alum solution cool down and transfer to a stoppered bottle. In this form the solution will remain active indefinitely and can be used each time skins have to be chromed. Paint the flesh side of the hides with the chrome solution, fold down the backbone skin to skin and store in a warm room for 12 hours. Repeat the painting application with the chrome solution and store for a further 12 hours. Find a thick section of the hide and cut with a knife to make sure that the chrome has penetrated the hide. An alternative way of checking is to place a small section of the chromed hide in boiling water, absence of puckering and shrinkage indicates that the hides have been adequately tanned. Rinse well in warm water, oil and strain.

OILING OR FAT LIQUORING OPERATION
Once the skins have been tanned by one of the above methods, the leather must be treated with an oil to preserve the suppleness and prevent them from becoming board-like. This can be achieved by preparing a 1:3 Neatsfoot oil/water emulsion and rubbing the skins between the fingers to work the oil into the grain. Having forced in as much oil as possible, fold the skins flesh to flesh and store for 2 days. Repeat this process and then proceed to the straining operation.

STRAINING AND DRYING TECHNIQUES
Up to this stage the leather has never been fully dried and with good reason. The skins should be dried on a frame or pegged out on a board and dried away from direct sunlight so that they will hold their shape over the years.

Chip board of suitable dimensions makes an ideal drying frame, although chicken wire tacked over a masonite frame will do the same job.

If you are using the chip board, take the partially dry hides and lightly tack, using carpet tacks near the edges of the leather. Do not overstretch or the contracting of the drying skins will tear the leather.

CARDING PROCESS

Now that you have produced your first leather sheepskin the only remaining process is to remove the grass seeds, long fibres and matted wool. Use a carding comb purchased from craft shops, or as an alternative, a steel comb. Comb out the burrs and seeds from the fleece. If necessary trim the fibres to an even length with a razor comb sold in hairdressers for trimming hair, or alternatively with a comb, tease out the fibres and trim to an even length with scissors or a razor blade. (This part is tricky and depends on patience and skill. It can be done beautifully and it is up to the student to experiment. It is the normal method of trimming sheep skins.)

**Dyeing the Fleece**

If you wish to colour the fleece it is a simple matter to dye the wool using disperse dyes at a temperature of 30°C. Make up a solution of the selected dyes and work the skins for one hour, depending on the shade required. Some of the best dyes are

|  |  |
|---|---|
| Dispersol range | I.C.I. Australia |
| Serilene Dyes | Yorkshire Chemical Ltd. |
| Cibacets | Ciba-Geigy |
| Artisils | Sandoz Aust. |

ALTERNATIVE TANNING PROCESS

There is a further sequence that can be applied to sheepskins and furs, and providing chemicals can be obtained from the Australian or New Zealand offices of Ciba-Geigy, the complete method is given in the following section. Although much of the information is the same as that which has been given before, the proprietary products are the same as commercial tanners are using. The results from these recipes are therefore the best available, although more expensive than in the previous section.

Liquor Ratio for all processes     20:1

*Soaking:* Cold water at 25°C with

|  |  |
|---|---|
| Tinovetin JU | 0.5 g/litre |
| Reversan 83 | 0.1 g/litre |

Time: 3 to 4 days.

*Fleshing* operation is as described in previous section.

*Scouring:* Cold water at 20°C with

|  |  |
|---|---|
| Sodium Carbonate | 1.5 g/litre |
| Tinovetin JU | 0.5 g/litre |

Scour for one hour, then rinse in running water.
*Washing:* Water at 25°C with

| | |
|---|---|
| Common salt | 10 g/litre |
| Ultravon AN | 1.5 g/litre |

Wash for 2 hours, then drain liquor.
*Bleaching:* Water at 30°C with

| | |
|---|---|
| Common salt | 30 g/litre |
| Clarite PS | 4-5 g/litre |
| Uvitex WGS | 0.5 g/litre |

Check pH to about 5: Treat for 1 hour.
*Pickling and Tanning:* Water at 30°C with

| | | |
|---|---|---|
| | Common salt | 60 g/litre 30 minutes |
| then add | Formic acid | 2 g/litre 30 minutes |

Leave in cooling bath and stir from time to time. Total time in bath about 24 hours. This bath is then used as the basis for the tanning operation.
*Tanning* (in the same bath as pickling)
Method A
To the Formic/common salt bath used above, add:

| | | |
|---|---|---|
| | Invasol 3909 | 3 g/litre 20 minutes |
| add | Tannesco H | 20 g/litre 30 minutes |

Leave standing for 20 minutes, stir for 30 minutes then stand for 2 hours.
**OR**
Method B:
To the Formic/common salt bath used in pickling, add:

| | | |
|---|---|---|
| | Potassium alum | 10 g/litre 20 minutes |
| then add | Tannesco H | 10 g/litre |

Stir 20 minutes, leave standing 2 hours, stir again 20 minutes and leave standing for further 2 hours.
*Basifying:* Water at 30°C with

Sodium Bicarbonate 2 g/litre 2 hours

Rinse skins under warm running water (not above 25°C) for 2 hours.
*Dyeing:* Water at 60°C with

| | | |
|---|---|---|
| | Acetic acid 60% | 1% |
| | Invaderm A | 1% stir for 10 minutes |
| then add | Invasol 3909 | 2% |
| | Erio type wool dyes | x% stir 30 minutes |

*Breaking:* Fold skin in half, flesh to flesh for 12 hours, let drain dry to partial dryness, then stake out to dry.

All chemicals and auxiliary assistants available from Ciba-Geigy.

# Appendix

## Chemical and Dyestuff Supply Sources

*Albright & Wilson*
610 St. Kilda Road, Melbourne. Vic. 3001
269 Grange Road, Findon. S.A. 5023
Levelling Agents, Calgon, Fabric Softeners, Detergents
*Ajax Chemicals*
Short Street, Auburn. N.S.W. 2144
Hamlet Street, Cheltenham. Vic. 3192
Chemicals for mordants and dyeing operations.
*BASF Australia*
55 Flemington Road, Nth Melbourne. 3051
Dyes, levelling agents, auxiliary agents etc.
*Bayer Australia*
47-67 Wilson Street, Botany. N.S.W. 2019
633-647 Springvale Road, Glen Waverley. Vic. 3150
Dyes, levelling agents, auxiliary agents etc.
*Robert Bryce*
145-147 Glenlyon Road, Brunswick. Vic. 3056
Chemicals, dyestuffs, auxiliary agents
*John Charlton & Co. Pty. Ltd.*
168-170 Pacific Highway, St Leonards, N.S.W. 2065
Tanning kits and supplies
*Ciba-Geigy Australia Ltd.*
P.O. Box 9 Northland Centre, Vic. 3072
P.O. Box 76 Lane Cove, N.S.W. 2068
Dyes, auxiliary agents, detergents etc.
Tanning chemicals and auxiliaries
*Croda Chemicals*
P.O. Box 1012 Richmond North, Vic. 3121
Chemicals, dyestuffs, auxiliary agents, detergents.
*Dye Services*
P.O. Box 54 Collingwood, Vic. 3066
Dyes, chemicals, auxiliary agents, detergents.

*Hodgsons Dye Agencies P/L*
26 William Street, Balaclava, Vic. 3183
Dyes, auxiliary agents, detergents.
*Hoechst, Australia Ltd.*
P.O. Box 25 Nunawading, Vic. 3132
Dyes, auxiliary agents, detergents.
*L.B. Holliday & Co. Ltd.*
8 Thornton Crescent, Mitcham, Vic. 3132
Dyes, auxiliary agents, detergents.
*ICI Australia, Petrochemicals Ltd.*
ICI House, 1 Nicholson Street, Melbourne, Vic. 3001
Dyes, chemicals, auxiliary agents, detergents.
*H.J. Langdon*
351-355 King Street, Melbourne. Vic. 3001
Chemicals for mordants and dyeing processes.
*Sandoz Australia*
P.O. Box 23 Chadstone, Vic. 3148
Dyes, chemicals, auxiliary agents, detergents
*H.B. Selby*
Ferntree Gully Road, Mt Waverley, Vic. 3149
Chemicals for mordants, glassware, thermometers.
*Tasmanian Craft Dyes & Chemicals Suppliers*
Box 256, P.O. Launceston, 7250
*Townson & Mercer*
200 Argyle Street, Hobart, Tas. 7000
Chemicals for mordants, dyeing processes, glassware, thermometers
*G.J. Vago & Sons P/L.*
58 Epsom Road, Rosebery, NSW 2018
Dyes, auxiliaries, detergents.
*A.E. Walker*
600 Burke Road, Camberwell, Vic. 3124
Chemicals for dyeing and mordants.
*Yorkshire Chemicals*
P.O. Box 209 Richmond, Vic. 3121
Dyes, auxiliary agents, detergents, tanning chemicals etc.

# Chemical and auxilliary index

| | |
|---|---|
| Acetic acid | 5, 24, 30, 32, 33, 37, 70, 80 |
| Albegal W | 31 |
| Albegal SW | 31 |
| Alum | see aluminium potassium sulphate |
| Aluminium acetate | 63 |
| Aluminium oxide | 55 |
| Aluminium potassium sulphate (alum, potassium alum) | 42, 45, 54, 76, 79 |
| Aluminium sulphate | 42, 43, 48 |
| Ammonia | 4, 6, 22, 33, 55, 58 |
| Ammonium sulphate | 27, 28, 31, 32, 72 |
| Basic aluminium sulphate | 48 |
| Basic chromium chloride | 42, 46 |
| Calcium acetate | 63, 64 |
| Calgon | 23, 24, 56 |
| Caustic soda (sodium hydroxide) | 21, 24, 35, 36, 37, 43, 47 |
| Chrome alum | 43, 47, 65, 77 |
| Chromium sulphate | 65 |
| Clarite P.S. | 79 |
| Common salt (sodium chloride) | 6, 34, 35, 36, 37, 73, 76, 77, 79 |
| Copper sulphate | 42, 45, 47, 64 |
| Creme of tartar | 42, 43, 44, 54, 55, 57 |
| Detergent A | 21, 24, 36, 49, 69, 75 |
| Detergent B | 23, 24 |
| Dispersol | 31 |
| Ferric sulphate | 43, 47, 65 |
| Ferrous sulphate | 42, 44, 65 |
| Formaldehyde | 77 |
| Formic acid | 6, 31, 33, 64, 76, 79 |

| | |
|---|---|
| Ground chalk | 63 |
| Hydrochloric acid | 57 |
| Hydrogen peroxide | 22, 23, 24 |
| Invaderm A | 80 |
| Invasol 3909 | 79, 80 |
| Lead acetate | 64 |
| Levelling agents | 6, 28, 31, 32, 33 |
| Lime solution | 65 |
| Lissapol NC | 36 |
| Manutex RS | 70 |
| Methylated spirits | 70, 72 |
| Mimosa extract | 76 |
| Neatsfoot oil | 77 |
| Nitric acid | 55 |
| Nobosol OT | 69 |
| Oxalic acid | 42, 44, 55, 59 |
| Paraffin wax | 69 |
| Potassium alum | see aluminium potassium sulphate |
| Potassium dichromate | 42, 44, 63, 64, 65 |
| Potassium ferricyanide | 66 |
| Potassium ferrocyanide | 66 |
| Reversan 83 | 78 |
| Soap | 17, 20, 33, 34, 37, 38 |
| Soda ash | see sodium carbonate |
| Sodium bicarbonate | 77, 79 |
| Sodium bisulphite | 65 |
| Sodium carbonate (soda ash, washing soda) | 6, 20, 23, 24, 35, 36, 43, 48, 63, 65, 77, 79 |

| | |
|---|---|
| Sodium chloride (see also common salt) | 34, 35, 36, 37 |
| Sodium dichromate | 33, 42, 44, 59, 65 |
| Sodium hydrosulphite | 35, 36, 37, 38 |
| Sodium hydroxide | see caustic soda |
| Sodium hypochlorite (White King) | 23, 24, 59 |
| Sodium metabisulphite | 23, 65, 73 |
| Sodium silicate | 23, 24, 42, 46 |
| Sodium sulphate | 5, 27, 30, 31, 32, 33, 34, 35 |
| Sodium sulphide | 35 |
| Sodium thiosulphate | 42, 45 |
| Softener | 56 |
| Stannic chloride | 42, 43, 45, 47, 64 |
| Stannous chloride | 42, 44, 57, 59, 63, 65, 66 |
| Sulphonated oil | 77 |
| Sulphuric acid | 6, 33, 58, 60, 66 |
| Swimming pool chlorine powder | 73 |
| | |
| Tannesco H | 79 |
| Tannic acid | 43, 46, 48, 62 |
| Tin metal | 55 |
| Tinovetin JU | 78, 79 |
| Tri sodium pyrophosphate | 22, 24 |
| Turkey red oil | 63 |
| | |
| Ultravon AN | 79 |
| Urea | 69 |
| Uvitex WGS | 79 |
| | |
| Washing soda (soda ash, sodium carbonate) | 6, 20, 23, 24, 35, 36, 43, 48, 63, 65, 77, 79 |
| Wetting agents | 69, 72, 73 |
| White King | 75 |

# General index

| | |
|---|---|
| Acid levelling dyes | |
| Nylon | 31, 69 |
| Silk | 31 |
| Wool | 31, 67 |
| Acid milling dyes | |
| Nylon | 32 |
| Silk | 32 |
| Wool | 32 |
| Acidity of dyebath | 6 |
| Air oxidation: vat dyes | 35 |
| Alizarin | |
| cotton dyeing | 62 |
| wool dyeing | 63 |
| Alkaline: chrome mordant | |
| cotton, flax | 46 |
| Aluminium mordant | |
| application | 42 |
| cotton | 48 |
| flax | 48 |
| silk | 45 |
| wool | 43 |
| Alum: skin tanning | 76 |
| Ammoniacal cochineal | 55 |
| Animal fibres: classification | 12 |
| Ardil | 13 |
| Auxilliary agents | 28 |
| Azoic dyes: cotton, flax | 34 |
| | |
| Barwood: wool dyeing | 62 |
| Basic concepts | 2, 27 |
| acidity | 6 |
| alkalinity | 6 |
| concentration | 6 |
| dyeing vessels | 7 |
| glassware | 9 |
| liquor ratio | 3, 28 |
| percentage calculations | 4, 28 |
| safety with chemicals | 10 |
| temperature | 3 |
| weights and volumes | 3 |

| | |
|---|---|
| Basifying chromed skins | 79 |
| Batik dyeing | 69 |
| Baumé | 47 |
| Bleaching recipes | |
| wool | 22 |
| silk | 22 |
| cotton | 23 |
| flax | 23 |
| tanned skins | 79 |
| Blind dyeing | 25 |
| Blow-out dyeing | 72 |
| Boil off liquor: silk dyeing | 34 |
| Brazil wood: wool dyeing | 62 |
| Breaking: tanned skins | 80 |
| Bush tanning skins | 76 |
| | |
| Calculations | |
| concentration | 6, 28 |
| depth of shade | 28 |
| liquor ratio | 3, 27, 43 |
| percentage | 4, 28, 43 |
| Camwood: wool dyeing | 62 |
| Carding and combing: skins | 78 |
| Cellulose | |
| acetate | 13, 38 |
| regenerated | 38 |
| Chemical solutions: labelling | 27 |
| Chrome mordants | |
| wool | 44 |
| silk | 46 |
| cotton | 47 |
| flax | 47 |
| Chrome dyes | |
| wool | 33, 44 |
| Chrome green | |
| wool | 65 |
| cotton | 65 |
| Chrome yellow | |
| cotton dyeing | 64 |
| Classification of fibres | 12 |

| | |
|---|---|
| Cochineal | |
|   crimson method | 54 |
|   one bath process | 55 |
|   rose pink process | 55 |
|   separate mordant process | 54 |
|   silver | 53 |
| Cockspur vine | 59 |
| Cold pad batch process | |
|   cotton | 72 |
|   wool | 69 |
| Cold vat dyeing process: | |
|   cotton | 37 |
| Colour lakes | 42 |
|   triangle: preparation | 40 |
| Concentration calculations | 6 |
| Copper mordant application | 42 |
|   wool | 45 |
|   silk | 45 |
|   cotton | 47 |
|   flax | 47 |
| Cotton: bleaching | 23 |
|   dyeing | |
|     azoic | 34 |
|     directs | 34 |
|     reactive | 34 |
|     soluble vats | 35 |
|     vat dyes | 35 |
|     mineral | 64 |
|     mordants | 46 |
|     mordant application | 42 |
|       aluminium | 43, 45, 48 |
|       chrome | 44, 46, 47 |
|       copper | 45, 47 |
|       iron | 44, 46, 47 |
|       tannic acid | 46 |
|       tin | 44, 45, 46 |
|     scouring | 21 |
| Cuba wood: wool dyeing | 59 |
| Cudbear: wool dyeing | 58 |

| | |
|---|---|
| Dacron | 13 |
| Degumming of silk | 21 |
| Delustered silk | 21, 33 |
| Depth of shade calculations | 28 |
| Dip dyeing | 67 |
| Direct dyes | |
|   cotton | 34 |
|   dip dyeing | 67 |
|   nylon | 34 |
|   silk | 34 |
|   tie dyeing | 69 |
|   viscose rayon | 38 |
| Disperse dyes | |
|   cellulose acetate | 38 |
|   nylon | 38 |
| Dissolving caustic soda | 47 |
| Dralon | 13 |
| Dyeing vessel construction | 7 |
| Dyeing operation | 27 |
| Dyeing theory | 2 |
| Dyeing tanned skins | 78 |
| Dyestuffs nomenclature codes | 30, 35 |
| | |
| Extraction of vegetable colouring matter | |
|   bark | 52 |
|   berries | 52 |
|   flowers | 52 |
|   husks | 52 |
|   leaves | 52 |
|   lichen | 58 |
|   moss | 58 |
|   nuts | 52 |
|   seed pods | 52 |
|   roots | 52 |
|   weeds | 52 |
| Extraction of tanning from | |
|   bark | 76 |

| | | | |
|---|---|---|---|
| Fastness, general | 13 | Jute | 12 |
| testing: light | 14 | | |
| rubbing | 17 | Kermes: wool dyeing | 57 |
| washing | 17 | | |
| Fat liquoring tanned skins | 77 | Labelling of dye and chemical | |
| Fermentation: lichen and moss | 58 | solutions | 27 |
| Flavine: wool dyeing | 60 | Lac: wool dyeing | 57 |
| Flax | 12, 46 | Lake formation | 9, 42 |
| mordant application | 46 | Lanolin: wool grease | 19 |
| aluminium | 48 | Levelness of dyed material | 25 |
| chrome | 47 | Levelling agents | 31 |
| copper | 47 | Lichen | 57 |
| iron | 47 | Light fastness testing | 14 |
| tannic acid | 46 | Light exposure mordant | |
| tin | 46 | treated wool | 44 |
| Fleshing of skins | 75 | Lima wood: wool dyeing | 62 |
| Formaldehyde tanning recipes | 77 | Liquor ratio | 3, 27, 43 |
| Fustic: wool dyeing | 59 | Logwood | |
| | | cotton dyeing | 64 |
| Glassware | 9 | wool | 64 |
| Grams | 3 | Lousy silk | 21 |
| Grams per litre (g/litre) | 3 | | |
| Greasy wool | 19 | Mädder: cotton dyeing | 63 |
| Green skins: tanning | 74 | wool dyeing | 62 |
| | | Man-made fibres | 12 |
| Haematine crystals | 64 | Man-made fibres: scouring | 22 |
| paste | 64 | Metallic salts | 42 |
| Hydrogen peroxide bleach | | Methylated spirits | 70 |
| cotton | 22 | Millilitres (ml) | 3, 28 |
| wool | 22 | Mineral colours | |
| Hydrometer | 47 | chrome green cotton | 65 |
| | | chrome green wool | 65 |
| Indigo: cotton dyeing | 37 | chrome yellow cotton | 64 |
| Iron buff: cotton dyeing | 65 | iron buff cotton | 65 |
| Iron mordant recipes | | mineral khaki cotton | 65 |
| cotton | 47 | Prussian Blue cotton | 65 |
| flax | 47 | wool | 66 |
| silk | 46 | Mineral fibres | 12 |
| wool | 44 | Mimosa extract tanning | 76 |

Mordant application
- wool
  - aluminium 43
  - chrome 44
  - copper 45
  - iron 44
  - tin 44
- *silk*
  - aluminium 45
  - chrome 46
  - copper 45
  - iron 46
  - tin 45
- *cotton and flax*
  - aluminium 48
  - chrome 47
  - copper 47
  - iron 47
  - tannic acid 46
  - tin 46
- process general 2, 42

Moss colouring matter 57

Natural dyes 50
Natural fibres 12
Neatsfoot Oil: tanning 77
Nylon classification 13
Nylon dyeing
- acid levelling 31
- acid milling 32
- chrome 33
- direct 34
- disperse 38
- premetallized 32
- reactives 33
- scouring 22

Oiling of tanned skins 77
Old fustic 59
Orlon 13

Orchella weed wool dyeing 57
Orchil wool dyeing 57

Oxidation of vat dyes 37
Oxygen in dyebath 37

pH papers 7, 21, 33
   scale 6
Pattern cards: 40
Peach wood wool dyeing 62
Pectins 21
Percentage calculations 4
Persian berries: wool dyeing 61
Pickling baths: tanning 76
Pipettes 9
Premetallized dyes
- nylon 32
- wool 30

Preparation dye and chemical solutions 4
Preparation natural colour baths 51
Prussian Blue
- cotton dyeing 65
- wool dyeing 66

Pyrex glass containers 8

Quercitron bark: wool dyeing 60

Rate of temperature rise:
- dyebaths 45

Rayon
- acetate 13, 38
- viscose 13, 38

Reactive dyes
- cotton 34
- wool 30

Red woods: wool dyeing 62
Regenerated cellulose 38
Rhus cotinus: wool dyeing 61
Rubbing fastness 17

| | |
|---|---|
| Safety with chemicals | 10, 43 |
| Safflower wool dyeing | 61 |
| Saffron wool dyeing | 61 |
| Salting sheep skins | 74 |
| Sapan wood wool dyeing | 62 |
| Saunders wood wool dyeing | 62 |
| Scouring recipes | |
|   cotton | 21 |
|   flax | 21 |
|   man-made fibres | 22 |
|   silk | 21 |
|   skins | 75 |
|   wool | 19 |
| Shellfish: colouring matter | 50 |
| Shrink resist treatment for wool | 73 |
| Silk fibre | 13 |
| Silk | |
|   bleaching | 22 |
|   dyeing | 29 |
|   degumming | 21 |
|   mordanting | 45 |
|   scouring | 21 |
| Silk mordant application | |
|   aluminium | 45 |
|   chrome | 46 |
|   copper | 45 |
|   iron | 46 |
|   tin | 45 |
| Silver cochineal wool dyeing | 53 |
| Skin | |
|   bleaching | 75 |
|   breaking | 75 |
|   carding | 78 |
|   dyeing | 78 |
|   fat liquoring | 77 |
|   oiling | 77 |
|   pickling | 76 |
|   soaking | 75 |
|   scouring | 75 |
|   straining | 78 |
|   tanning | 76 |

| | |
|---|---|
| Soap powder | 20 |
| Soaking of skins | 75 |
| Soluble vat dyes: cotton | 35 |
| Solution preparation | |
|   dyes | 4, 25 |
|   chemicals | 4, 27 |
| Softening: dyed and mordanted wool | 19, 44, 52 |
| Stainless steel equipment | 9 |
| Standard formula | |
|   powders | 4, 25 |
|   undiluted chemicals | 4, 27 |
|   diluted dyes | 5, 28 |
|   diluted chemicals | 5, 28 |
| Stirring rods | 9, 43 |
| Straining of tanned skins | 78 |
| Suint | 19 |
| Sulphur dyes: cotton | 35 |
| Suppliers: dyes and chemicals | 81 |
| Swimming pool chlorine powder | 73 |
| Synthetic fibres: classification | 12 |
| Tanning skins recipes | |
|   alum | 76 |
|   chrome | 77 |
|   formaldehyde | 77 |
|   vegetable | 76 |
|   industrial | 79 |
| Temperature | |
|   control | 3, 75 |
|   rate of rise | 3 |
|   scale | 3 |
| Terylene | 13 |
| Testing | |
|   light | 14 |
|   rubbing | 17 |
|   washing | 17 |
| Tie dyeing | 69 |

| | | | | |
|---|---|---|---|---|
| Tin mordant | | | Wool | |
| cotton | 46 | | bleaching | 22 |
| flax | 46 | | dyeing | |
| silk | 45 | | acid levelling | 29 |
| wool | 44 | | acid milling | 29 |
| Turkey red oil process | 48, 63 | | chrome | 30 |
| | | | cochineal | 54 |
| Urea | 69 | | dye woods | 59 |
| | | | flavine | 60 |
| Valparaiso weed: wool dyeing | 57 | | fustic | 59 |
| Vat dyes application | | | kermes | 57 |
| cotton | 36 | | lac | 57 |
| viscose rayon | 38 | | lichen | 57 |
| Vegetable dyes general | 52 | | mineral colours | 64 |
| Vegetable fibres: classification | 12 | | moss | 57 |
| Vegetable tanning: skins | 76 | | persian berries | 61 |
| Viscose rayon | 13, 38 | | premetallized | 30 |
| Volumes | 3 | | Prussian Blue | 66 |
| | | | quercitron | 60 |
| Washing fastness | 17 | | reactives | 30 |
| skins | 75 | | rhus | 61 |
| Water absorbancy: cotton | 17 | | safflower | 61 |
| Weights: measures | 3 | | saffron | 61 |
| Weld: wool dyeing | 61 | | weld | 61 |
| Wetting agents | 69 | | Wool grease | 19 |
| Wooden spoons | 43 | | mordant application | |
| | | | aluminium | 43 |
| | | | chrome | 44 |
| | | | copper | 45 |
| | | | iron | 44 |
| | | | tin | 44 |
| | | | scouring | 19 |
| | | | Yellow wood: wool dyeing | 59 |